Quantum Circuit Simulation

George F. Viamontes • Igor L. Markov • John P. Hayes

Quantum Circuit Simulation

Springer

George Viamontes
Department of Electrical Engineering &
Computer Science
Advanced Computer Architecture Laboratory
gviamont@eecs.umich.edu
http://www.eecs.umich.edu/~gviamont/

Igor L. Markov
Department of Electrical Engineering &
Computer Science
University of Michigan
Ann Arbor, MI 48109-2121
USA
imarkov@eecs.umich.edu

John P. Hayes
Department of Electrical Engineering &
Computer Science
University of Michigan
Ann Arbor, MI 48109-2121
USA
jhayes@eecs.umich.edu

ISBN 978-90-481-3064-1 e-ISBN 978-90-481-3065-8
Springer Dordrecht Heidelberg London New York

Library of Congress Control Number: 2009929020

Printed on acid-free paper

Springer is part of Springer Science+Business Media (www.springer.com)

Preface

Recent scientific advances, both experimental and algorithmic, have brought quantum information processing closer to reality. The ability to process information "quantumly" already allows ultra-precise metrology and more secure communication, and quantum algorithms allow computations such as factoring to be done significantly faster than we know how to do them on classical computers. With these advances, quantum simulation has become an increasingly important topic for both theoretical and engineering reasons. On the theoretical front, progress toward defining the class of circuits that can be simulated efficiently on a classical computer has and will continue to lead to a deeper understanding of the power of quantum computation. Even though efficient simulation of all quantum circuits may not be possible, the circuits that will most likely make up the majority of operations on a quantum computer can in fact be simulated efficiently. These are fault-tolerant error-correction circuits; they are composed of only Clifford-group gates, which, as this book demonstrates, can be simulated surprisingly efficiently on a classical computer.

Recent results suggest that larger classes of circuits can also be simulated efficiently on classical computers. From a circuit perspective, efficient simulation can result from operating with a restricted set of gates, such as the Clifford-group gates, or from operating with an arbitrary gate set if the circuit has a small treewidth, as Markov and Shi have shown. From a physical perspective, efficient simulation can also result from limiting the amount of entanglement in the intermediate states of the computation. For example, Vidal has shown that one-dimensional many-body systems can be simulated efficiently and more recently, Bravyi and Terhal have identified a large class of quantum Hamiltonians for which the adiabatic evolution can be simulated efficiently on a classical computer. Ideally, we would like to characterize the class of quantum circuits that are computationally no more efficient than the equivalent classical versions. Such developments would lead to new theoretical results, sharpening our understanding of the differences between classical complexity classes and their quantum counterparts. It would also help us understand the true power of quantum computation.

On the engineering front, efficient simulation provides the ability to validate and perform sanity checks on circuit components prior to their implementation. It is also useful for evaluating circuits, in particular for benchmarking the error rates of gates and memory. Through simulation, engineers can save on both the cost and effort involved in building hardware. Quantum simulation can also offer software support for quantum control and algorithm design.

The main requirement of a quantum simulator is that its output match the information produced by an ideal quantum computer. It may also produce additional information, such as descriptions of states before measurement. A quantum circuit can be simulated efficiently only when the input and output can be expressed in a compact form, since in general a quantum circuit may require an exponential-size matrix to express the evolution and an exponentially long vector to express the state. Thus, a quantum simulator should be able to operate on limited amounts of entanglement, to represent (and possibly compress) quantum evolution and states in a classical data structure, and to perform classical algorithms. The QuIDDPro simulator described in this book offers precisely this functionality.

This book is a major step forward in the development of algorithms and software tools for designing and simulating quantum circuits. Using a consistent notation, it describes in detail simulation techniques that up to this point had been scattered throughout the research literature in physics, computer science, and computer engineering journals.

The book describes an innovative software system, called QuIDDPro, that can be used to describe and simulate quantum circuits. The QuIDDPro software is freely available and can be used as a "quantum calculator" by researchers interested in developing new quantum algorithms and by students interested in learning quantum information processing. QuIDDPro has already been used by students in quantum computing courses to learn the behavior of quantum circuits. The book also describes algorithms and techniques that will be indispensable to computer scientists and engineers developing languages, compilers, optimizers, and CAD tools for quantum computation.

The algorithmic toolbox implemented in QuIDDPro is based on a powerful new data structure called a quantum information decision diagram (QuIDD). QuIDDs are the secret behind QuIDDPro's ability to simulate important classes of quantum circuits with remarkable efficiency. QuIDDPro simulations of important classes of quantum circuits can be several orders of magnitude faster than similar computations done with common numerical software tools such as Matlab. The book describes in detail the underlying mathematical models, algorithms, and data structures that are used to implement QuIDDPro. Two appendices introduce the reader to the QuIDDPro software and show how to implement some common quantum algorithms in QuIDDPro, including Grover's search algorithm and Shor's integer factoring algorithm.

The book is self contained, covering the fundamentals of linear algebra and quantum mechanics needed to understand quantum circuits and to use the simulator. It requires only basic familiarity with algebra, graph algorithms, and computer engineering.

The experience of the authors that combines quantum computing research with electronic design automation has yielded a book that is a must read for anyone interested in the simulation of quantum algorithms and in the implementation of design tools for quantum computers.

Alfred V. Aho
Columbia University, New York, NY

Krysta M. Svore
Microsoft Research, Redmond, WA

Contents

1 Introduction 1
1.1 Quantum Circuits 1
1.2 Quantum Simulation 3
1.3 Book Outline 4

2 Gate Modeling and Circuit Simulation 7
2.1 Classical Digital Circuits 7
2.2 Simulation with Binary Decision Diagrams 11
2.3 Sequential Circuits and Synchronization 17
2.4 Summary 18

3 Linear Algebra and Quantum Mechanics 19
3.1 Linear Algebra 19
3.2 Quantum Mechanics 24
3.3 Summary 32

4 Quantum Information Processing 33
4.1 Quantum Gates 33
4.2 Quantum Circuits 38
4.3 Synchronization of Quantum Circuits 42
4.4 Sample Algorithms 43
4.5 Summary 46

5 Special Case: Simulating Stabilizer Circuits 47
5.1 Basics of a Quantum Circuit Simulator 47
5.2 Stabilizer States, Gates and Circuits 49
5.3 Data Structures 51
5.4 Algorithms 52
5.5 Summary 55

6 Generic Circuit Simulation Techniques 59
6.1 Qubit-wise Multiplication 59
6.2 P-blocked Simulation 61
6.3 Tensor Networks 63
6.4 Slightly-entangled Simulation 66
6.5 Summary 70

7 State-Vector Simulation with Decision Diagrams 71
7.1 Quantum Information Decision Diagrams 71
7.2 Scalability of QuIDD-based Simulation 80
7.3 Empirical Validation 88
7.4 Related Decision Diagrams 92
7.5 Summary 100

8 Density-Matrix Simulation with QuIDDs 103
8.1 QuIDD Properties and Density Matrices 103
8.2 QuIDD-based Outer Product 105
8.3 QuIDD-based Partial Trace 106
8.4 Empirical Validation 109
8.5 Summary 114

9 Checking Equivalence of States and Circuits 115
9.1 Quantum Equivalence Checking 115
9.2 Global-Phase Equivalence 117
9.3 Relative-Phase Equivalence 122
9.4 Empirical Validation 126
9.5 Summary 128

10 Improving QuIDD-based Simulation 133
10.1 Gate Algorithms 133
10.2 Dynamic Tensor Products and Partial Tracing 139
10.3 Empirical Validation 145
10.4 Summary 150

11 Closing Remarks 153

A QuIDDPro Simulator 155
A.1 Running the Simulator 155
A.2 Functions and Code in Multiple Files 158
A.3 Language Reference 160

B QuIDDPro Examples 177
B.1 Well-known Quantum States 177
B.2 Grover's Search Algorithm 178
B.3 Shor's Integer Factoring Algorithm 179

References 181

Index 187

1

Introduction

The construction of computer algorithms and software models that simulate physical systems plays a fundamental role in all branches of science and engineering. The physicist and Nobel laureate Richard Feynman, among others, observed in the 1980s that the important task of simulating quantum-mechanical processes on a standard computer requires an extraordinary amount of computer memory and runtime [41]. Such observations gave rise to the notion of *quantum computing*, where quantum mechanics itself is used to simulate quantum behavior. The key insight is to replace the familiar 0 and 1 bits of conventional or *classical* computing with information units called qubits (quantum bits) that capture quantum states of elementary particles or atomic nuclei. By operating on qubits, a quantum computer can, in principle, process exponentially more data than a classical computer in a similar number of steps. In the 1990s, several fast quantum methods were discovered for such applications as searching large databases [38] and factoring large numbers [82]; the latter is a basic step in some forms of codebreaking.

1.1 Quantum Circuits

Implementing quantum algorithms in physical hardware, that is, building quantum computing circuits, has proven to be extremely difficult. A method called liquid-state nuclear magnetic resonance (NMR) was used around 2001 to demonstrate Shor's factoring algorithm running on a 7-qubit quantum computer [103]. Atoms in an organic molecule were used to represent individual qubits, and radio-frequency (RF) pulses were used to address the qubits. However, this technique is not scalable to greater numbers of qubits, in large part because of the practical difficulties of addressing individual qubits in big molecules. Significant progress was also reported in the 2000s using several unrelated implementation techniques, but most still suffer from scalability limitations.

A particularly promising class of quantum circuit technologies rely on semiconductor devices, and so can draw on the massive investment already made by the semiconductor industry in reliable and scalable chip manufacturing techniques for very large-scale integrated (VLSI) circuits. A quantum dot [104] is a particle—a single electron, for example—that is trapped in a semiconductor in a way that enables it to represent a qubit. Quantum dots can be fabricated in gallium arsenide (GaAs) chips and addressed individually by means of conventional wires. Ion traps [69], originally developed at the U.S. National Institute of Standards and Technology (NIST) for atomic clocks, use carefully engineered electric fields to suspend electrically charged atoms (ions) of cesium or beryllium in long chains, with vibrational coupling between them. These ions also represent individual qubits and can be addressed by lasers. While the first laboratory demonstrations of ion traps used very bulky equipment, more recent ion traps reside entirely within tiny GaAs chips. A third semiconductor technology relies on superconducting qubits that are implemented with Josephson junctions. Despite its need for an extremely low-temperature (cryogenic) environment, this technology promises to significantly reduce power consumption in large-scale quantum circuits, and has also been proposed for conventional super-computers. Prototypes of superconducting quantum circuits were built in a number of laboratories using variants of the technology.

Quantum information processing by means of quantum circuits and algorithms has already proven practical in secure optical communications, where photons serve as the qubits. Quantum communication has several distinct advantages over conventional methods. For instance, in classical communication, eavesdroppers can escape detection. When qubits rather than bits are being communicated, on the other hand, eavesdropping is readily detected due to the fact that it requires a quantum-mechanical measurement that alters the state of the measured information. Moreover, quantum states cannot be copied, therefore quantum messages cannot easily be saved for future decryption. In 2007, demonstrations of single-photon quantum communication in optical fiber and free space achieved ranges in excess of 100 kilometers.

Currently, fiber-based quantum communication technologies are being commercialized by several companies in the US and Europe. Their potential customers include financial institutions, military organizations, and governments. For example, the results of the October 2007 general election in Geneva, Switzerland were transmitted using quantum communications. Another application that is being actively explored is quantum communication between satellites and ground stations, where the transmission medium is inherently insecure, and physical contact between the communicating parties is impractical. As with traditional communication technologies, increasingly sophisticated quantum communication protocols and networks must be supported by quantum information-processing circuits, whose simulation is the central topic of this book.

1.2 Quantum Simulation

Software simulation has long been an invaluable tool for the design and testing of classical systems, such as electrical circuits, digital logic circuits, communication systems, etc. [39]. In the digital circuit domain, which is most relevant to this work, simulation is typically considered a computer-aided design (CAD) task, and was itself once thought to be computationally intractable. Naive simulation and synthesis techniques for n-bit digital circuits require $\Omega(2^n)$ runtime and memory, that is, the computational complexity of these techniques tends to grow exponentially with the circuit size n. Later advances in algorithm design brought about the ability to perform circuit simulation far more efficiently in many practical cases. One such advance was the development of a data structure called the *reduced ordered binary decision diagram* (ROBDD) [20], which greatly compresses large collections of digital signals, and allows direct manipulation of the compressed form. Similarly, software simulation can be expected to play a vital role in the development of quantum computing hardware by enabling the modeling and analysis of large-scale designs before they are implemented physically.

The mathematics needed for simulating quantum processes, including quantum computational algorithms, is the linear algebra of complex-valued vector spaces. Such processes can often be modeled by quantum circuits, which are analogous to classical digital circuits. Unfortunately, straightforward simulation of quantum circuits by classical computers executing standard linear-algebraic routines requires $\Omega(2^n)$ time and memory [41, 61].[1] However, just as ROBDDs and other innovations have made the simulation of very large classical circuits tractable, clever algorithmic techniques can allow the efficient simulation of quantum circuits in many important cases. Interestingly, if a classical computer can simulate a quantum circuit or algorithm solving a particular problem, then this implies that a classical computer is computationally as powerful as a quantum computer for the problem in question. Therefore, by discovering new classical algorithms which can efficiently simulate quantum algorithms in certain cases, we are probing the limitations of quantum computing.

One unavoidable limitation of quantum computing is that it is extremely error-prone due to the fact that, unlike classical bits, qubits interact with their environment in ways that cause their values to decay as time passes. Another is that reading the value of a qubit—quantum measurement—is a non-deterministic process, which gives quantum information processing a

---

[1] Quantum circuits can be simulated using polynomial-sized memory resources, but the dramatic increase in runtime required by such techniques to periodically recompute intermediate results makes them impractical. The best known complexity-theoretic results are discussed in [32]. In particular, quantum circuit simulation belongs to the complexity class P^{PP} which includes decision problems solvable in polynomial time with the help of an oracle for solving problems from PP, i.e., those solvable in probabilistic polynomial time.

probabilistic character quite different from most classical computation. In this work, we describe the development of practical software methods for simulating general quantum circuits. Such simulation can be used as a tool to address the following problems:

1. Characterizing the effect of errors in physical quantum circuits
2. Evaluating error-correction techniques to cope with such errors
3. Verifying the correctness of synthesized quantum circuits
4. Exploring the boundaries between quantum and classical computation

A number of algorithms may form the basis for a quantum simulator Such algorithms typically rely on some type of structure present in quantum circuits or intermediate quantum states. We illustrate these issues by describing the theoretical framework of several existing quantum algorithms. A particular goal of the book is to demonstrate how to take any of these theoretical methods and develop a practical software tool to address the analysis and synthesis issues for quantum circuits. Several chapters focus on a particular data structure, the quantum information decision diagram (QuIDD), that we developed in our research [95, 97, 99]. QuIDD-based simulation differs from most other techniques in that it automatically detects many types of useful structures in the target circuits. For example, it is surprisingly effective at simulating instances of Grover's algorithm, both theoretically and empirically. This book presents and proves some key results about QuIDD-based simulation. It also provides an in-depth look at the implementation of a QuIDD-based simulator, QuIDDPro, to demonstrate the design and evaluation of a complete simulation package.

Several other quantum simulation techniques have been developed that excel under different circumstances. For example, an important class of quantum circuits known as stabilizer circuits can be simulated very efficiently with the specialized algorithms of Gottesman, Knill and Aaronson [36, 1]. A broad range of "slightly-entangled" circuits can be simulated by a technique due to Vidal [106, 107], while circuits implementing the frequently-used Quantum Fourier Transform can be approximately simulated by the tensor-contraction technique of Markov and Shi [55], as pointed out by Aharonov et al. [4] as well as by Yoran and Short [112]. The book discusses all of these techniques, although in less detail than QuIDD-based simulation.

1.3 Book Outline

Chapter 2 reviews circuit models used in digital logic, as well as current methods of simulating and verifying traditional logic circuits. A key data structure for representing and manipulating logic functions, the binary decision diagram (BDD) and its applications, is discussed. For readers who are less familiar with the relevant mathematical and quantum mechanical background, Chapter 3 reviews the basic concepts relevant to the topics covered in this book. Chapter

4 introduces the quantum circuit model with particular emphasis on connections to its linear-algebraic underpinnings.

Chapter 5 outlines the major features that every quantum circuit simulator must implement using a well-known simulation technique called the stabilizer formalism as an example. Chapter 6 expands this discussion by describing a variety of other recently developed computational techniques. The quantum information decision diagram (QuIDD) and the simulator QuIDD-Pro are presented in detail in Chapter 7 and expanded upon in Chapter 8. QuIDDPro implements the QuIDD data structure and all related algorithms with an expressive front-end language. This chapter also discusses a practical class of quantum states and operators that can be simulated efficiently using QuIDDs, with quantum search serving as the main benchmark algorithm for evaluation purposes. Chapter 9 addresses the verification of synthesized quantum circuits by simulating the circuits and checking for equivalence among the resultant states and operators. We show in Chapter 10 that QuIDDPro's input language enables some automatic speed-up techniques for QuIDD-based simulation. Finally, Chapter 11 summarizes the trends of quantum information science addressed in the book, and gives some perspectives on possible future applications. Appendix A introduces the reader to the QuIDDPro software that implements many of the algorithms covered in this book. Appendix B presents QuIDDPro source code for several well-known quantum circuits.

Acknowledgments

Our work on this book was partially supported by the DARPA QuIST program, US AirForce Research Laboratory (agreement No. FA8750-05-1-0282) and National Science Foundation (award No. 0208959). We are also thankful to colleagues and students who contributed valuable comments at various stages of our work, including Al Aho, Héctor Garcia, Smita Krishnaswamy, Manoj Rajagopalan and Yaoyun Shi. A number of users of our quantum circuit simulator QuIDDPro reported bugs and made valuable suggestions, which helped us improve the software. Full documentation and latest versions are going to be available at `http://vlsicad.eecs.umich.edu/Quantum/qp/`

2

Gate Modeling and Circuit Simulation

As with other sophisticated technologies in information processing, the practical implementation and use of electronic circuits is preceded by their mathematical modeling and simulation. Detailed calculations of physical parameters are employed to characterize individual circuit elements with respect to environmental conditions and manufacturing variations, but are difficult to scale to very large integrated circuits. In this context, a compact model captures the essence of a circuit's operation without unnecessary details. Such models express the expected outputs of each circuit component given its inputs, and are key to fast algorithms for simulating the functional behavior of the entire circuit: its speed, power dissipation, temperature distribution, etc. Simulation is thus an important prerequisite to validating the operation of the circuit.

This chapter discusses the modeling and simulation of classical digital circuits to introduce and motivate the methods used for quantum circuit simulation. Emphasis is placed on functional simulation and its use in formal verification of both combinational and sequential circuits.

2.1 Classical Digital Circuits

The literature on modeling and simulating traditional electronic circuits is extensive, and is stratified by the level of detail sought. For example, the electrical behavior of wires, transistors and small circuits is typically modeled in terms of ordinary differential equations using the SPICE [85, 72] simulation tool and its variants. Reduced models are used in fast timing analysis [72, 71], while logic-level simulation that propagates 0s and 1s from a circuit's inputs to its outputs, requires even fewer computational resources because it can represent gates by simple look-up tables. Figure 2.1 illustrates the modeling of a standard inverter (NOT gate) at several different levels of abstraction.

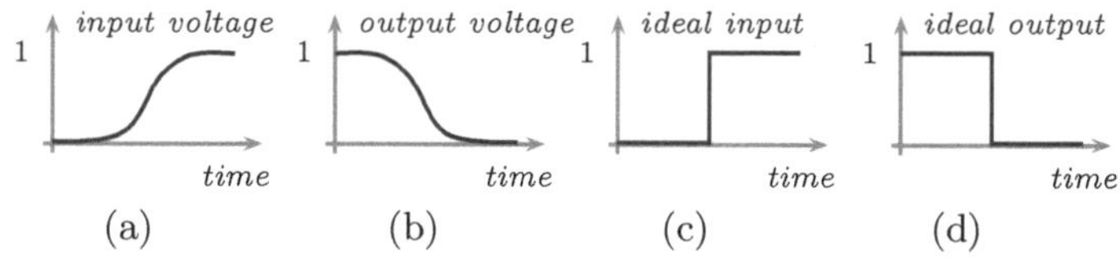

input	output
0	1
1	0

(e)

Fig. 2.1. Modeling inverter behavior at the electrical and logical levels: Analog voltage signals at (a) the input and (b) the output; digital voltage signals at (c) the input and (d) the output; (e) compact logic-level model (truth table).

Simulating Single Input Combinations

We illustrate logic-level simulation using the small circuit example in Figure 2.2a, whose function is expressed by the Boolean equation:

$$F(x, y, z) = \text{NAND}(\text{NAND}(x, y), \text{NOT}(z))$$

This is an example of a *combinational* logic circuit, meaning no memory elements are present and time is effectively ignored. Later, in Section 2.3, we will discuss the simulation of *sequential* circuits, which include memory and take time into account.

Standard simulators assume fixed input values, and traverse the circuit from primary inputs to primary outputs in topological order, implying that a gate is evaluated only after all of its predecessors have been evaluated. The output of every gate is looked up in a truth table, such as that of Figure 2.2b, based on the values available at the gate's inputs. These values may be produced by similar look-ups for predecessor gates. The result of such functional simulation on all possible input combinations is presented in the form of a truth table in Figure 2.2c, although simulation of larger circuits is often restricted to a subset of the entries in such tables.

The gains in simulation speed that result from the use of truth tables facilitate simulation algorithms whose runtime scales linearly with circuit size [39], allowing them to deal with circuits of any practical size. The same algorithmic template (topological traversal with table look-up) can be extended to timing simulation that estimates the duration of each signal transition [72]. A key idea is to approximately capture the waveforms from Figures 2.1a-b by a small set of parameters, for example, the logic-level transition, the timing of the 50%-point, or the slope (angle) of the transition. The look-up tables used in static analysis tools capture gate delay for all possible functional transitions and several values of continuous parameters, so that intermediate values can be handled by linear interpolation. It is interesting to note that the slope parameter is difficult to observe directly, but often affects the timing of downstream transitions.

Functional simulation can be repeated for different input combinations, and timing analysis can be performed for different input transitions. Since such invocations of the simulator are independent, they can easily be parallelized when the same circuit must be simulated many times.

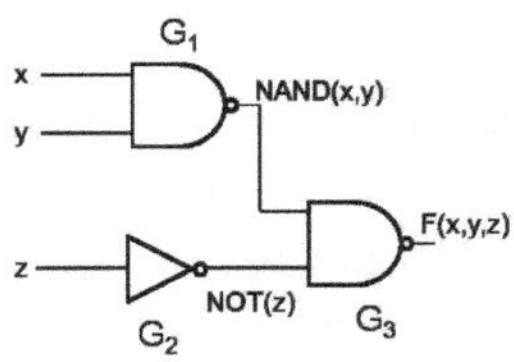

(a)

input		*output*
x	y	NAND
0	0	1
0	1	1
1	0	1
1	1	0

(b)

input			*output*
x	y	z	$F(x,y,z)$
0	0	0	0
0	0	1	1
0	1	0	0
0	1	1	1
1	0	0	0
1	0	1	1
1	1	0	1
1	1	1	1

(c)

Fig. 2.2. Simulating a small logic circuit: (a) graphical depiction, (b) truth table of a NAND gate, (c) simulated values at the circuit's output in truth-table form.

Simulating Multiple Input Combinations

Linear-time algorithms for logic and timing simulation achieve their efficiency by abstracting away unnecessary details. Further abstraction helps in simulating multiple input combinations or input transitions at once, which can be useful to compute the worst-case (critical) delays through the circuit, in order to determine the highest possible clock frequency. Logic simulation for multiple input combinations can determine if the circuit can ever produce incorrect results or enter a prohibited state. Extending timing analysis to estimate worst-case delays is relatively easy if the delay of every gate is represented by the worst possible case, regardless of the functional values at the gate's inputs. This pessimistic assumption facilitates a linear-time algorithm which slightly over-estimates the worst possible delay between any input transition and the stabilization of the circuit outputs [39].

Extending logic-level simulation to handle many input combinations at once is rather difficult. *Exhaustive simulation*, that is, trying all possible input combinations one by one, requires prohibitive amounts of time for circuits with more than about 30 inputs. *Symbolic simulation*, a technique developed in the 1980s, attempts to reuse and share computations normally performed for different input combinations. The term "symbolic" refers here to the fact that the information propagated through the circuit during a topological traversal does not consist of individual numeric parameters, but rather of graph-based data structures (symbols) that capture multiple parameter values through an abstract transformation. Such data structures are introduced in Section 2.2

and implicitly capture large sets of value combinations by means of specially-designed directed acyclic graphs [39].

Symbolic simulation starts by representing all input combinations by a rather straightforward graph data structure. It proceeds by (symbolically) applying individual gate operations to this data structure, which results in implicit representations of intermediate value combinations. The algorithms involved traverse and modify the graphs that encode these combinations. As a result, the data structures typically grow in size. After symbolically simulating all gates, one obtains a representation of all possible output value combinations. Dedicated algorithms can then verify if prohibited combinations appear on the outputs. This technique can be quite helpful for small- and medium-sized circuits in that it saves a great deal of runtime compared to exhaustive simulation. However, it often requires much more memory, and on large circuits gives rise to the so-called *memory explosion* problem, when memory requirements skyrocket after the data structure reaches a certain size. Symbolic simulation is also notoriously difficult to parallelize.

In one technique that is fundamental to both single-input and symbolic simulation, gate operations are not applied one by one, but are first conglomerated into clusters that are modeled by larger gates (this can be done once, but will speed up simulation of many inputs). This technique is illustrated in Figure 2.2 and removes the need to simulate signals within the clusters. It may, however, be limited by the size of look-up tables that represent these larger gates. To circumvent this limitation, symbolic simulation can represent the function of a larger gate implicitly (symbolically) by a specially-designed graph, as shown in later sections. Simulation algorithms are then extended to work with value-combination graphs and function graphs, rather than explicit look-up tables. Other possible extensions involve partitioning large graphs and performing symbolic simulation on one partition at a time.

Circuit Verification

Simulation can be used to verify the functional correctness of a given circuit. For example, if the circuit in question is an optimized variant of another circuit that is known to be correct, one can formulate an equivalence-checking problem to confirm that the two circuits always produce the same outputs when given identical inputs. If the two circuits are not equivalent, one seeks an input combination for which the outputs differ. Exhaustive simulation of all possible input combinations suffices when the number of inputs is small. To check the equivalence of two larger circuits, it is common to use a "miter" circuit, which connects the target circuits to the same input source and produces 0 when their outputs are identical and 1 when their outputs differ. The miter is formed by connecting the corresponding outputs of the circuits being compared to XOR gates; the outputs of the XORs are then combined by an OR gate. This construction reduces equivalence checking to the task of checking if the miter always produces 0, which can be addressed efficiently by symbolic simulation.

In later chapters, we will draw some parallels between the simulation of quantum and conventional logic circuits—simulating quantum superposition is similar to simulating multiple input combinations at the same time. To this end, we will extend symbolic simulation to the quantum case by viewing (square) matrices that capture quantum operators as analogous to the truth tables used to simulate conventional gates. On the other hand, important differences exist, which include the necessary use of complex-valued arithmetic when modeling quantum circuits, as well as several different notions of equivalence. Another major difference is that quantum circuits often exhibit probabilistic behavior, while conventional logic circuits are deterministic. As is the case with conventional electronic circuits, physically-accurate modeling is possible for quantum circuits based on partial differential equations (such as the Schrödinger equation), but such detailed simulation is beyond the scope of this book.

2.2 Simulation with Binary Decision Diagrams

This section describes an important data structure that has been found extremely useful for symbolically simulating classical logic circuits. The key property of this data structure is that it can represent many signal combinations in a compressed form which can be manipulated directly without decompression. Such compression makes it possible to represent exponentially many signal combinations in a format whose size is linear in the number of inputs. As we will see later, the ability to exponentially compress many data states in this way is especially important for the simulation and verification of quantum circuits. First, we need to understand the original data structure from which the quantum techniques are derived, namely the *binary decision diagram* (BDD).

Binary Decision Diagrams

The BDD was introduced by Lee in 1959 [51] in the context of classical logic circuit design. It represents a Boolean function $f(x_1, x_2, ..., x_n)$ by means of a directed acyclic graph (DAG); see Figure 2.3. By convention, the top node of a BDD is labeled with the name of the function f represented by the BDD. Each variable x_i of f is associated with one or more nodes that have two outgoing edges labeled *then* (solid line) and *else* (dashed line). The *then* edge of node x_i denotes an assignment of logic 1 to x_i, while the *else* edge represents an assignment of logic 0. These nodes are called *internal nodes* and are labeled by the corresponding variable x_i. The edges of the BDD point downward, implying a top-down assignment of values to the Boolean variables depicted by the internal nodes.

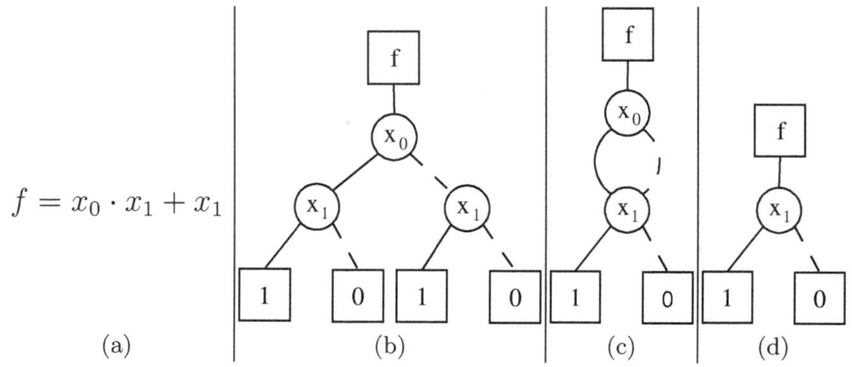

Fig. 2.3. (a) A logic function, (b) its BDD representation, (c) its BDD representation after applying the first reduction rule, and (d) its ROBDD representation.

At the bottom of the BDD are box-shaped *terminal nodes* containing the logic values 1 or 0. They denote the output value of the function f for a given assignment of its variables. Each path through the BDD from top to bottom represents a specific assignment of 0-1 values to the variables $x_1, x_2, ..., x_n$ of f, and ends with the corresponding output value $f(x_1, x_2, ..., x_n)$. For example, Figure 2.3b shows a BDD representing the two-valued Boolean function $f(x_0, x_1) = x_0 \cdot x_1 + x_1$. To determine, the value of $f(0, 1)$, one proceeds downward from the top node f to the internal node x_0. Then the outgoing *else*(dashed) branch from node x_0, corresponding to $x_0 = 0$ is taken to internal node x_1. Finally, the outgoing *then* (solid) branch from x_0 is traversed leading to a terminal node marked 1. One therefore concludes that $f(0, 1) = 1$.

The original BDD data structure conceived by Lee has exponential memory complexity $\Theta(2^n)$, where n is the number of Boolean variables in a given logic function. The reason for this complexity bound is that in Lee's initial design, the paths corresponding to all possible 2^n combinations of variable assignments are explicitly represented, as in Figure 2.3b. Moreover, exponential memory and runtime are required in many practical cases, making this data structure impractical for simulation of large logic circuits. To address this limitation, Bryant developed the *reduced ordered BDD* (ROBDD) [20], where all variables are ordered, and decisions are made in that order. A key advantage of the ROBDD is that variable ordering facilitates an efficient implementation of rules that automatically eliminate redundancy from the basic BDD representation. These *reduction rules* may be summarized as follows:

Reduction Rule 1. There are no nodes v and v' such that the subgraphs rooted at v and v' are isomorphic
Reduction Rule 2. There are no internal nodes with *then* and *else* edges that both point to the same node

An example of how the rules transform a BDD into an ROBDD is shown in Figure 2.3. The subgraphs rooted at the x_1 nodes in Figure 2.3b are isomorphic. By applying the first reduction rule, the BDD in Figure 2.3b is converted into the BDD in Figure 2.3c. Notice that in this new BDD, the *then* and *else* edges of the x_0 node now point to the same node. Applying the second reduction rule eliminates the x_0 node, producing the ROBDD in Figure 2.3d. Intuitively, it makes sense to eliminate the x_0 node since the output of the original function is determined solely by the value of x_1. An important aspect of redundancy elimination is the sensitivity of ROBDD size to the variable ordering. Finding the optimal variable ordering is an NP-hard problem, but efficient ordering heuristics have been developed for specific applications. Moreover, it turns out that many practical logic functions have ROBDD representations that are polynomial (or even linear) in the number of input variables [20]. In addition, the reduction rules make ROBDDs *canonical*, which means that no two different ROBDDs represent equivalent Boolean functions. Thus, equivalence of ROBDDs can be checked in $O(1)$ time by simply comparing the root nodes. Consequently, ROBDDs have become indispensable tools in the design, simulation, and synthesis of classical logic circuits.

Associated Algorithms

Even though the ROBDD is often quite compact, efficient algorithms are necessary to make it practical for logic circuit simulation. Thus, in addition to the foregoing reduction rules, Bryant introduced a variety of ROBDD operations whose complexities are bounded by the size of the ROBDDs being manipulated [20]. Of central importance is the **Apply** operation, which performs a binary operation with two ROBDDs, producing a third ROBDD as the result. It can be used, for example, to compute the logical AND of two logic functions. **Apply** is implemented by a recursive traversal of the two ROBDD operands. For each pair of nodes visited during the traversal, an internal node is added to the resultant ROBDD using the three **Apply** rules defined graphically in Figure 2.4. To understand these rules, some notation must be introduced. Let v_f denote an arbitrary node in an ROBDD f. If v_f is an internal node, $Var(v_f)$ is the Boolean variable represented by v_f, $T(v_f)$ is the node reached when traversing the *then* edge of v_f, and $E(v_f)$ is the node reached when traversing the *else* edge of v_f.

Clearly the **Apply** rules depend on the variable ordering. To illustrate, consider performing **Apply** using a binary operation *op* and two ROBDDs f and g. **Apply** takes as arguments two nodes, one from f and one from g, and the operation *op*. This is denoted as **Apply**(v_f, v_g, op). **Apply** compares $Var(v_f)$ and $Var(v_g)$ and adds a new internal node to the ROBDD result using its three rules. These rules also guide **Apply**'s traversal of the *then* and *else* edges (this is the recursive step). For example, suppose **Apply**(v_f, v_g, op) is called and $Var(v_f) \prec Var(v_g)$. Rule 1 is invoked, causing an internal node containing $Var(v_f)$ to be added to the resulting ROBDD.

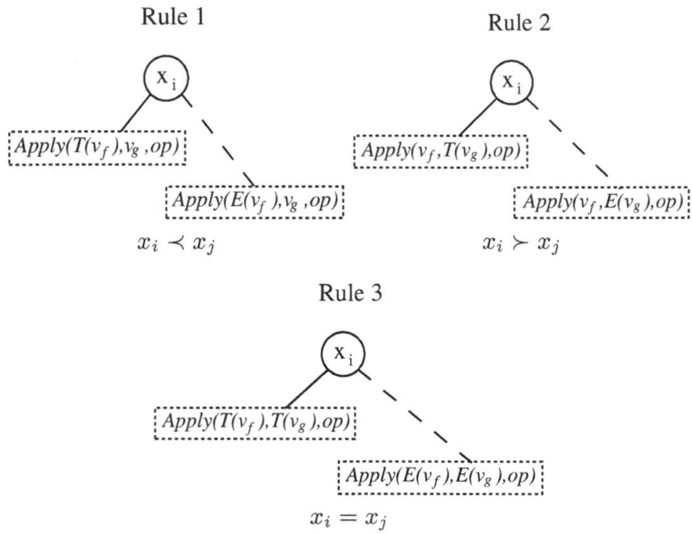

Fig. 2.4. The three recursive rules used by the **Apply** operation that determine how a new node should be added to an ROBDD. Here $x_i = Var(v_f)$ and $x_j = Var(v_g)$. The notation $x_i \prec x_j$ means that x_i precedes x_j in the variable ordering.

Apply Rule 1 then directs the **Apply** operation to call itself recursively with **Apply**$(T(v_f), v_g, op)$ and **Apply**$(E(v_f), v_g, op)$. **Apply** Rules 2 and 3 dictate similar actions but handle the cases when $Var(v_f) \succ Var(v_g)$ and $Var(v_f) = Var(v_g)$. To recurse over both ROBDD operands correctly, the initial call to **Apply** must be **Apply**$(Root(f), Root(g), op)$ where $Root(f)$ and $Root(g)$ are the root nodes for the ROBDDs f and g.

The recursion stops when both v_f and v_g are terminal nodes. When this occurs, *op* is performed with the values of the terminals as operands, and the resulting value is added to the current ROBDD as a terminal node. For example, if v_f contains the value logical 1, v_g contains the value logical 0, and op is defined to be $\oplus$ (XOR), then a new terminal with value $1 \oplus 0 = 1$ is added to the ROBDD result. Terminal nodes are considered *after* all variables are considered. Thus, when a terminal node is compared to an internal node, either Rule 1 or Rule 2 will be invoked depending on which ROBDD the internal node is from. The pseudocode for **Apply** is provided in Figure 2.5. The one-operand (unary) version is very similar.

Figure 2.6 demonstrates ROBDDs and **Apply** for the circuit depicted earlier in Figure 2.2. The ordering used in this example is $x \prec y \prec z$. First $NAND(x, y) = \mathbf{Apply(x, y, NAND)}$ is computed. This involves the application of Rule 1 on x and y followed by Rule 3 on the terminal nodes. The recursion ends when applying $NAND$ to the terminal values of x and y (all terminal nodes are last in the variable ordering). Subsequently, $F(x, y, z) = \mathbf{Apply(NAND(x, y), NOT(z), NAND)}$ is computed. This step

```
Apply(A, B, op) {
  if (Is_Constant(A) and Is_Constant(B)) {
    return New_Terminal(op(Value(A),
      Value(B)))
  }
  if (Table_Lookup(R, op, A, B)) return R
  v = Top_Var(A, B)
  T = Apply(A_v, B_v, op)
  E = Apply(A_v', B_v', op)
  R = ITE(v, T, E)
  Table_Insert(R, op, A, B)
  return R
}
```

Fig. 2.5. Pseudocode for the `Apply` algorithm. `Top_Var` returns the variable index from either `A` or `B` appearing earlier in the ordering, while `ITE` creates a new internal node with children `T` and `E`.

involves applying **Apply** Rule 1 twice by comparing x and y with z in sequence. **Apply** Rule 3 terminates the recursion by applying $NAND$ to the terminal values of $NAND(x, y)$ and z.

Some simple examples of logical reductions automatically generated by **Apply** are worth mentioning. First, notice that in the intermediate result, $NAND(x, y)$, if x is set to 0 (the dashed edge), then the result of the whole function is immediately determined without having to evaluate any assignments to y. This optimization is reflected by the fact that the dashed edge of x is connected to a 1 terminal node without any connection to a y node. Similarly, in the final result $F(x, y, z)$, when $NAND(x, y)$ is set to 0 (i.e., both x and y are set to 1), the outgoing solid edge of y is directly connected to a 1 terminal node with no need to evaluate any assignments to z. Notice however, that another potential optimization when $NOT(z)$ is set to 0 is overlooked (i.e., the solid outgoing edge of z). If the variable ordering had instead been $z \prec x \prec y$, then this optimization would be present in the final ROBDD result, skipping any evaluation of x and y. This important point highlights the fact that finding a good variable ordering is critical when trying to maximize the compression gained from an ROBDD. Although finding an optimal ordering for a given function is, in general, NP-hard as noted earlier, heuristics exist which can often find near-optimal orderings at low computational cost, thereby compressing many functions in practice [83].

The success of ROBDDs in making a seemingly difficult computational problem tractable in practice has led to the development of numerous ROBDD variants outside the domain of logic design. Of particular relevance to this work are multi-terminal binary decision diagrams (MTBDDs) [25] and algebraic decision diagrams (ADDs) [6]. These data structures are compressed representations of matrices and vectors rather than logic functions, and the

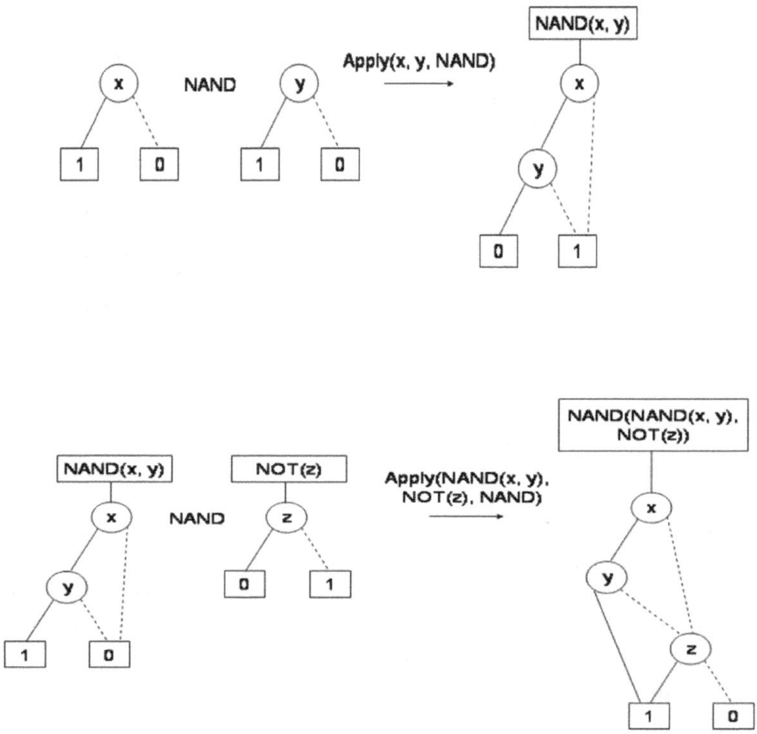

Fig. 2.6. ROBDD and **Apply** example using the circuit in Figure 2.2; complemented edges are not shown. The edge attributes are required for ROBDDs to be canonical; their use also improves compression [39].

amount of compression achieved is proportional to the frequency of repeated values in a given matrix or vector. Additionally, some standard linear-algebraic operations, such as matrix multiplication, are defined for MTBDDs and ADDs. Since these operations are based on **Apply**, their efficiency is proportional to the size in nodes of the MTBDDs or ADDs being manipulated. Further discussion of the MTBDD and ADD representations is deferred to Chapter 7 where the general structure of the QuIDD is described.

2.3 Sequential Circuits and Synchronization

Modern digital circuits, ranging from room thermostats to laptop computers, can contain millions of gates and perform millions of operations per second. The time-dependent or *sequential* behavior of the circuits—when the signals being processed interact with one another and how they are kept in step—is a very important design issue.

The overwhelming majority of digital circuits are classified as *synchronous sequential circuits*, which means that their computations are broken into sequences of small steps each requiring a fixed unit of time called a *clock period* or *clock cycle* T [40]. At the end of each clock period, newly computed results are collected so they can be used as inputs at start of the next clock period. A special control signal called the *clock* is employed to define the clock period and is distributed as needed to all parts of the circuit. In each clock period, the clock signal is 1 for some fixed time and then switches to 0. Thus the clock oscillates back and forth between 0 and 1 in a uniform and precisely timed manner. The frequency $f = 1/T$ in cycles per second (hertz) is a basic measure of circuit speed. A laptop might have a clock frequency of 1 gigahertz ($f = 10^9$ hertz) implying a clock period of 1 nanosecond ($T = 1$ ns).

Synchronization by means of a clock greatly simplifies the circuit design process by relieving the designer of the need to do detailed, on-the-fly signal scheduling. It is only necessary to ensure that, in each clock period, all signals assume their correct final values before a well-defined deadline determined by the clock. Reliability is also increased by building a margin of safety into the clock period that allows for minor variations in the times at which signals change. The latter include fluctuations (noise) in the delays though signal transmission paths, which are often not known accurately because they depend on uncontrollable aspects of the circuit's manufacturing process or operating conditions. While *asynchronous* circuits in which no clock is present have potential performance advantages in some situations, these are outweighed by dramatic complications in design verification and testing. As we shall see in later chapters, quantum circuits must sometimes be synchronized in an even more rigorous way than classical circuits.

Synchronous sequential circuits can be partitioned into several parts: combinational sub-circuits (those we have discussed so far), separated by memory elements that are controlled by a common clock signal. The memory is typically composed of 1-bit storage elements called *D flip-flops* which operate as follows. When the clock signal switches from 0 to 1, the flip-flop samples the signal applied to its data input D and stores the resulting 0 or 1 value Q until the next rising edge of the clock appears [40]. A flip-flop thus can supply a newly computed Q signal to the circuit once each clock period.

The stored value Q is called the current *state* of the flip-flop, and all the flip-flops combine to express the state of the entire circuit. The possible states of the circuit can be enumerated and represented by vertices of a graph. For each signal combination that can appear on the circuit's external inputs, we can establish an arrow (directed edge) connecting the current state to the next state. This mathematical description of a sequential circuit is called a finite-state machine (FSM) or a finite-state automaton (FSA) [39]. An FSM model thus captures the computational behavior of a circuit and allows one to estimate the time (in clock cycles) needed to complete a computation. A similar notion of a quantum finite automaton (QFA) is reviewed in Section 4.3.

Many of the concepts discussed in the previous sections extend to synchronous sequential circuits. A sequential circuit can be simulated by decomposing its operation into one-cycle steps, and simulating its combinational logic once per clock period. The results computed in each period are stored and used for (combinational) simulation in the next clock period; this works for both single-input and multiple-input simulation.

To verify the equivalence of two sequential circuits, we can first try to verify their combinational parts, e.g., using ROBDDs. If they are equivalent and connected in exactly the same way, we conclude that the entire sequential circuits in question are equivalent [39]. However, it is more likely that sequential circuits believed equivalent have different combinational parts, and this kind of equivalence is much harder to check. While we can verify equivalence over some given number of clock periods, the first behavioral differences may appear many cycles later. Consequently, sequential verification remains an open research problem, but much of the software used to verify commercial microprocessors uses ROBDDs for this task.

2.4 Summary

This chapter recounted the important role played by simulation techniques in digital circuit design, and reviewed the data representations, in particular, the binary decision diagram (BDD), used in symbolic simulation of both combinational and sequential circuits. Decision diagrams will also play a key role in Chapter 7 when we explore operations with compressed matrices and the QuIDD data structure.

With a working knowledge of classical circuit simulation in place, we now move to Chapter 3, which presents the basic mathematical concepts required to understand quantum circuit simulation.

3

Linear Algebra and Quantum Mechanics

Boolean algebra is commonly viewed as the mathematical foundation of classical logic circuits. The analogous role for quantum logic circuits is played by linear algebra. The physical principles of quantum circuits are governed by quantum mechanics, which makes heavy use of the theory of linear operators in Hilbert space. Section 3.1 reviews the portions of linear algebra that are most relevant to quantum mechanics. Then Section 3.2 reviews the key concepts of quantum mechanics that are most relevant to quantum circuits.

3.1 Linear Algebra

A quantum computation consists of a sequence of operations performed on quantum states. The operations and states can be represented by matrices and vectors, respectively, and are required by the underlying physics (quantum mechanics) to follow the rules of linear algebra. This section reviews the basic algebraic concepts we need.

Matrices and Vectors

A straight line in the 2-dimensional plane is described by an equation of the form

$$a_1x_1 + a_2x_2 = b$$

This generalizes to a set of m *linear* equations in the n-dimensional space formed by n variables $x_1, x_2, \ldots, x_n$:

$$\begin{aligned} a_{11}x_1 + a_{12}x_2 + \cdots + a_{1n}x_n &= b_1 \\ a_{21}x_1 + a_{22}x_2 + \cdots + a_{2n}x_n &= b_2 \\ &\vdots \\ a_{m1}x_1 + a_{m2}x_2 + \cdots + a_{mn}x_n &= b_m \end{aligned} \tag{3.1}$$

These equations may be written in matrix form as

$$AX = B \tag{3.2}$$

Here A denotes an $m \times n$ matrix, while X and B denote (column) vectors, which are also $n \times 1$ matrices.

$$A = \begin{bmatrix} a_{11} & a_{12} & \cdots & a_{1n} \\ a_{21} & a_{22} & \cdots & a_{2n} \\ \vdots & \vdots & \vdots & \vdots \\ a_{m1} & a_{m2} & \cdots & a_{mn} \end{bmatrix} \quad X = \begin{bmatrix} x_1 \\ x_2 \\ x_3 \\ \vdots \\ x_n \end{bmatrix} \quad B = \begin{bmatrix} b_1 \\ b_2 \\ b_3 \\ \vdots \\ b_n \end{bmatrix}$$

We are interested in the case where the matrix and vector entries are complex numbers and the set of vectors forms a *complex vector space* of n dimensions, denoted C^n; here C is the set of complex numbers. An example of a vector in C^2 is $V = \begin{bmatrix} 1.3 + 6i \\ 2.5 \end{bmatrix}$ where as usual $i = \sqrt{-1}$. Every complex number z may be written in the form $a + bi$, where a and b are real numbers. Associated with z is another complex number called its *complex conjugate* $z^* = a - bi$. If we multiply z by z^* we obtain

$$(a + bi)(a - bi) = a^2 + b^2$$

which is a real number. The quantity $\sqrt{a^2 + b^2}$ is known as the *modulus* of z and is a measure of its size.

Consider the following n vectors in the n-dimensional vector space C^n:

$$B_1 = \begin{bmatrix} 1 \\ 0 \\ 0 \\ \vdots \\ 0 \end{bmatrix} \quad B_2 = \begin{bmatrix} 0 \\ 1 \\ 0 \\ \vdots \\ 0 \end{bmatrix} \quad \cdots \quad B_n = \begin{bmatrix} 0 \\ 0 \\ 0 \\ \vdots \\ 1 \end{bmatrix} \tag{3.3}$$

The vectors $B_1, B_2, \ldots, B_n$ form a *basis* for C^n, meaning that any vector V in C^n can be expressed as a linear combination of the basis vectors. For example, for the vector V in C^2 we can write

$$V = \begin{bmatrix} 1.3 + 6i \\ 2.5 \end{bmatrix} = (1.3 + 6i) \begin{bmatrix} 1 \\ 0 \end{bmatrix} + 2.5 \begin{bmatrix} 0 \\ 1 \end{bmatrix} = (1.3 + 6i)B_1 + 2.5B_2$$

More generally, a basis for C^n is any set of n vectors that are linearly independent and span the entire vector space in the sense that every vector can be expressed as a linear combination of the basis vectors.

Matrix Operations

There are many ways to transform a given matrix or vector into a new one with useful properties. For example, every matrix M has a *transpose* M^T, which is formed by interchanging the rows and columns of M. The transpose of the 2×1 column vector $V = \begin{bmatrix} 1.3+6i \\ 2.5 \end{bmatrix}$ is the 1×2 row vector $V^T = \begin{bmatrix} 1.3+6i & 2.5 \end{bmatrix}$. A matrix can also be mapped into a single (complex) number or *scalar* in various ways. For instance, the *trace* $Tr(M)$ of an $n \times n$ or *square* matrix M is the sum of the entries along its main diagonal. Thus, the trace of $M = \begin{bmatrix} 1+i & 0 \\ 7 & 2.14+i \end{bmatrix}$ is $(1+i)+(2.14+i) = 3.14+2i$.

Two matrices can be combined in various ways. The overall effect of several consecutive matrix operations acting on the same system is obtained my multiplying the matrices. Multiplying an $m \times n$ matrix M by an $n \times p$ matrix N results in the (ordinary) *product* matrix P denoted $M \cdot N$ or, more simply, MN whose dimensions are $m \times p$. For example,

$$M = \begin{bmatrix} 1+i & 0 \\ 7 & -i \end{bmatrix} \quad N = \begin{bmatrix} 1-i & 0 & 9 \\ i & i & 0 \end{bmatrix} \quad P = MN = \begin{bmatrix} 2 & 0 & 9+9i \\ 7-7i & 1+i & 63 \end{bmatrix}$$

Equations 3.1 an 3.2 illustrate the product of a matrix by a vector. In general, if the entry in row i and column j of M is denoted by M_{ij}, the matrix product $P = MN$ is defined by

$$P_{ij} = \sum_{k=1}^{n} M_{ik} N_{kj} \tag{3.4}$$

It is possible to apply the foregoing product operation to vectors in several ways. Of interest in the context of quantum mechanics, is the *inner product* of a $1 \times n$ row vector X by an $n \times 1$ column vector, which is a scalar defined by

$$\sum_{i=1}^{n} X_i^* Y_i = [X_1^* X_2^* \ldots X_n^*] \begin{bmatrix} Y_1 \\ Y_2 \\ \vdots \\ Y_n \end{bmatrix} \tag{3.5}$$

Observe that the entries in the left vector of the product are the complex conjugates of the entries in X. In quantum mechanics, a complex-valued vector space with an inner product is called a *Hilbert space*. The inner product indicates the similarity between two vectors. If the inner product is zero, the vectors are said to be *orthogonal.*

It is easy to see that the product YX of a row vector by a column vector, which is known as the *outer product,* is an $n \times n$ square matrix. The outer product of a vector by itself is a *projection* matrix. Consider $P_i = B_i B_i^T$ where B_i is the ith basis vector defined in Equation 3.3. For the 3-dimensional case,

$$P_1 = \begin{bmatrix} 1\,0\,0 \\ 0\,0\,0 \\ 0\,0\,0 \end{bmatrix} \quad P_2 = \begin{bmatrix} 0\,0\,0 \\ 0\,1\,0 \\ 0\,0\,0 \end{bmatrix} \quad P_3 = \begin{bmatrix} 0\,0\,0 \\ 0\,0\,0 \\ 0\,0\,1 \end{bmatrix}$$

The effect of multiplying a column vector V by the projection matrix P_i is to make all entries in V zero, except the ith entry, which is unchanged.

Another special type of square matrix which is often encountered is the $n \times n$ *identity* matrix denoted by I or I_n. It is defined by the property that $MI = IM = M$ for any matrix M. It follows from Equation 3.4 that I has the following form with 1s on the main diagonal and 0s everywhere else:

$$I_n = \begin{bmatrix} 1\,0\,0 \cdots 0 \\ 0\,1\,0 \cdots 0 \\ \vdots \quad \ddots \quad \vdots \\ 0\,0\,0 \cdots 1 \end{bmatrix}$$

If N is such that $MN = NM = I$, then N is the *inverse* of M. The inverse matrix is usually written as M^{-1}. To illustrate,

$$M = \begin{bmatrix} 1 & 2 \\ 1 & 3 \end{bmatrix} \qquad M^{-1} = \begin{bmatrix} 3 & -2 \\ -1 & 1 \end{bmatrix}$$

Clearly I is its own inverse, but not every matrix has an inverse.

Another matrix operation of special interest in quantum computation is the *tensor product* (sometimes referred to as the Kronecker product) denoted by $M \otimes N$. It provides the basic way for combining two or more (quantum) systems into one larger one. If M and N are $m \times n$ and $p \times q$ matrices, respectively, then $M \otimes N$ is an $mp \times nq$ matrix formed by multiplying every entry M_{ij} by N, thus replacing M_{ij} by the submatrix $M_{ij}N$. For example,

$$I_2 \otimes V = \begin{bmatrix} V & 0 \\ 0 & V \end{bmatrix} = \begin{bmatrix} 1.3+6i & 0 \\ 2.5 & 0 \\ 0 & 1.3+6i \\ 0 & 2.5 \end{bmatrix}$$

In the case of the identity matrix, we have

$$I_m \otimes I_n = I_{mn}$$

Observe that $I_2 \otimes V \neq V \otimes I_2$, so the tensor product, like the matrix product, is not, in general, commutative.

Suppose A is a matrix with complex entries. The *adjoint* or *complex conjugate* of A denoted A^* is the matrix formed by first forming A's transpose A^T, and then replacing every entry in A^T by its complex conjugate. For example,

$$A = \begin{bmatrix} 1+2i & -i \\ 4.65 & -7+i \\ 0 & 2.75-i \end{bmatrix} \quad A^* = \begin{bmatrix} 1-2i & 4.65 & 0 \\ i & -7-i & 2.75+i \end{bmatrix}$$

A matrix which is its own adjoint, that is, $A = A^*$, is said to be *Hermitian.* A matrix whose inverse is the same as its adjoint is said to be *unitary.* In other words, A is unitary if $A^{-1} = A^*$. Three well-known matrices in quantum mechanics are the 2×2 *Pauli matrices*:

$$\sigma_x = \begin{bmatrix} 0 & 1 \\ 1 & 0 \end{bmatrix} \quad \sigma_y = \begin{bmatrix} 0 & -i \\ i & 0 \end{bmatrix} \quad \sigma_z = \begin{bmatrix} 1 & 0 \\ 0 & -1 \end{bmatrix} \tag{3.6}$$

It can easily be seen that each of these matrices is its both its own adjoint and its own inverse, that is, $\sigma_i = \sigma_i^* = \sigma_i^{-1}$. Hence, the Pauli matrices are both Hermitian and unitary.

Eigenvectors and Eigenvalues

Suppose that A is an $n \times n$ (square) matrix. A vector X is an *eigenvector* of A if AX is a scalar multiple of X, that is, there is a scalar λ called an *eigenvalue* of A such that

$$AX = \lambda X$$

In the case of the Pauli matrix σ_y of Equation 3.6, we have

$$\begin{bmatrix} 0 & -i \\ i & 0 \end{bmatrix} \begin{bmatrix} 1 \\ -i \end{bmatrix} = \begin{bmatrix} -1 \\ i \end{bmatrix} = -1 \begin{bmatrix} 1 \\ -i \end{bmatrix}$$

Hence, $\begin{bmatrix} 1 \\ -i \end{bmatrix}$ is an eigenvector of σ_y with eigenvalue -1. It is easily verified that σ_y has a second eigenvector $\begin{bmatrix} 1 \\ i \end{bmatrix}$ with eigenvalue $+1$.

In quantum mechanics, a Hermitian matrix A corresponds to some measurable quantity or *observable.* Its eigenvalues $a_1, a_2, \ldots, a_n$ are real numbers, that is, they have no imaginary parts, reflecting the fact that measurement outcomes must be real numbers. The corresponding eigenvectors $A_1, A_2, \ldots, A_n$ form a basis in the relevant Hilbert space, so an arbitrary vector can be completely defined in terms of the A's eigenvectors and eigenvalues. A itself can also be completely characterized by its eigenvectors and eigenvalues as follows. Let $P_i = A_i A_i^*$ be a projection matrix corresponding to eigenvector A_i and eigenvalue a_i. Then

$$A = a_i \sum_{i=1}^{n} P_i$$

This result is known as the *spectral decomposition* of A. Observe that the identity matrix I_n can be written as the spectral decomposition $\sum_{i=1}^{n} P_i$.

3.2 Quantum Mechanics

The physics underlying quantum computing is quantum mechanics, so a brief overview of the important properties of quantum mechanics is in order. Although there is more than one model of quantum mechanics, we restrict the presentation to the *Dirac* model which makes extensive use of linear algebra. This model applies to infinite-dimensional vector spaces, but the finite-dimensional case suffices for our purposes.

Fundamental Postulates

Quantum mechanics, and therefore quantum computing, is governed by four fundamental postulates that have been verified over the years by many experiments. Any simulation of quantum computing must implement these four postulates in some form if true quantum behavior is to be modeled. A brief summary of the four postulates follows; a more detailed discussion can be found in a number of standard quantum-mechanical texts [61, 74].

Postulate 1. Quantum states are represented as vectors in a Hilbert space.

Thus the space of quantum states is a complex-valued vector space for which the inner product (Equation 3.5) is defined. For our purposes, this means that qubits, which are quantum states, are represented as vectors for which we can compute inner products. The need for a vector representation comes from a physical phenomenon called superposition of states. In quantum computing devices, two low-energy stable states are used to represent the classical values 0 and 1, and are referred to as *computational basis states* [61]. Like an analog signal, the range of qubit values is an infinite continuum of values between 0 and 1. However, unlike an analog signal, these values denote a *probability* of obtaining a 0 or 1 upon measurement of a qubit. This is the essence of superposition.

It is convenient at this point to adopt the bra-ket notation for vectors and vector operations invented by Paul Dirac [74]. The symbol $|X\rangle$ denotes a *ket* in the standard *Dirac notation* and is short-hand for a complex-valued column vector. The row vector X^*, which is the adjoint of X, is denoted by the symbol $\langle X|$ called a *bra*. The matrix-vector product $M \cdot X$ can be written as $M|X\rangle$. The inner product of X by Y defined by Equation 3.5 is then the (scalar) product $\langle X| \cdot |Y\rangle$, which is denoted concisely by $\langle X|Y\rangle$. If we use $|i\rangle$ to denote the ith basis vector B_i defined by Equation 3.3, we can use Dirac notation to pick out any element M_{ij} of a given matrix M thus: $M_{ij} = \langle i| M |j\rangle$.

Intuitively, a qubit ψ has two components: α representing the amount of "zeroness" and β representing the amount of "oneness" that the qubit contains. We can combine these into a column vector and represent the result by a ket thus:

$$|\psi\rangle = \begin{bmatrix} \alpha \\ \beta \end{bmatrix}$$

More formally, α and β are complex numbers called *probability amplitudes*. The corresponding squares $|\alpha|^2$ and $|\beta|^2$ are the probabilities of measuring the qubit as a 0 and as a 1, respectively. Since a set of probabilities must sum to one, Postulate 1 requires that $|\alpha|^2 + |\beta|^2 = 1$.

In an equal or balanced superposition, $|\alpha|^2 = |\beta|^2$, and a qubit in such a state is interpreted as being both 0 and 1 simultaneously. Mathematically, an equal superposition of one qubit has the form $|+\rangle = \begin{bmatrix} 1/\sqrt{2} \\ 1/\sqrt{2} \end{bmatrix}$. It is easy to see that $|\frac{1}{\sqrt{2}}|^2 + |\frac{1}{\sqrt{2}}|^2 = 1$. The states $|0\rangle$ and $|1\rangle$ have the vector forms $\begin{bmatrix} 1 \\ 0 \end{bmatrix}$ and $\begin{bmatrix} 0 \\ 1 \end{bmatrix}$, respectively. Clearly, they form a basis for a two-dimensional complex vector space. Thus $|0\rangle$ and $|1\rangle$ are called *basis states*; they may be equated with the classical bits 0 and 1, respectively.

As will be demonstrated shortly, the massive parallelism achieved in quantum computing is due largely to both the property of superposition and Postulate 4. Furthermore, since the state vectors associated with qubits exist in a finite-dimensional Hilbert space, their inner products must be defined. This property is shown to be important in both Postulate 3 and the no-cloning theorem, which are discussed later.

Postulate 2. Operations on quantum states in a closed system are represented using matrix-vector multiplication of the quantum state vector by a unitary matrix.

This postulate suggests that unitary matrices are analogous to logic gates in classical computation. We can therefore view unitary matrices as *operators* that can modify the values of qubits just as logic gates modify classical bits. For the remainder of this book, the terms operator and gate are used interchangeably.

Unlike classical logic gates, however, all quantum operators are reversible, a consequence of their unitarity. Given a sequence of operations performed on a set of qubits, the qubits can be returned to their original state simply by performing the inverse of each operation in the reverse order that the operations are applied. Mathematically speaking, suppose we want to modify an initial qubit state $|\psi\rangle$ using the unitary matrices A, B, and C to produce a new state $|\psi'\rangle$. This is accomplished through a series of multiplications: $ABC\,|\psi\rangle = |\psi'\rangle$. Then $|\psi\rangle$ is recovered by multiplying in reverse order the inverse of each of the matrices by the new state: $C^{-1}B^{-1}A^{-1}\,|\psi'\rangle = |\psi\rangle$.

Since all quantum operators must be unitary, there exists an inverse for every operator and that inverse is the adjoint of the operator. Thus, by keeping track of the operations performed on a set of qubits, any quantum computation can be reversed by applying the adjoint of each operation in reverse order.

An example of a commonly used operator in quantum computing is the Hadamard operator, which has the matrix form

$$H = \begin{bmatrix} \frac{1}{\sqrt{2}} & \frac{1}{\sqrt{2}} \\ \frac{1}{\sqrt{2}} & -\frac{1}{\sqrt{2}} \end{bmatrix}$$

This operator is frequently used to put a qubit into an equal superposition state, as described in Postulate 1. To illustrate, we can transform a qubit in state $|0\rangle$ into an equal superposition via matrix-vector multiplication using the Hadamard operator as follows:

$$\begin{bmatrix} \frac{1}{\sqrt{2}} & \frac{1}{\sqrt{2}} \\ \frac{1}{\sqrt{2}} & -\frac{1}{\sqrt{2}} \end{bmatrix} \begin{bmatrix} 1 \\ 0 \end{bmatrix} = \begin{bmatrix} \frac{1}{\sqrt{2}} \\ \frac{1}{\sqrt{2}} \end{bmatrix}$$

Postulate 3. Measurement of a quantum state $|\psi\rangle$ involves a special set of operators. When any such operator Ω is applied to $|\psi\rangle$, the result will be one of the eigenvalues ω of Ω with a certain probability. Measurement is destructive and will change the measured state $|\psi\rangle$ to $|\omega\rangle$.

In the context of quantum computing, this postulate has two major consequences. The first is that measuring the value of a qubit destroys its quantum state, forcing it to a discrete 0 or 1 value corresponding to classical bit states, which are not superpositions of values. After measurement is performed, a quantum computation is no longer reversible in the strict sense implied by Postulate 2. The second consequence is that measurement is probabilistic.

There are several different types of quantum measurement, but the one that is most pertinent to this discussion is *measurement in the computational basis.* In quantum computing, this means measuring with respect to the $|0\rangle$ or $|1\rangle$ basis states of a qubit, and thereby forcing the qubit to a classical 0 or 1 state. The actual outcome (i.e., a 0 or 1 measurement) depends on the probability amplitude associated with each outcome in the superposition of the qubit (defined as α and β in Postulate 1). In making a measurement Ω_x of this type on quantum state ψ, the probability $p(x)$ of obtaining a 0 or 1 corresponds to the product $\langle\psi| \Omega_x |\psi\rangle$ where x is either 0 or 1.

Suppose, for example, we want to measure $|\psi\rangle = \begin{bmatrix} \alpha \\ \beta \end{bmatrix}$ in the $|1\rangle$ basis. The operator for this "projective" measurement is the projection matrix $P_1 = |1\rangle\langle 1| = \begin{bmatrix} 0 & 0 \\ 0 & 1 \end{bmatrix}$ (compare Equation 3.6). Consequently, the probability $p(1)$ of observing a 1 is

$$p(1) = \langle\psi|1\rangle\langle 1|\psi\rangle = \begin{bmatrix} \alpha^* & \beta^* \end{bmatrix} \begin{bmatrix} 0 & 0 \\ 0 & 1 \end{bmatrix} \begin{bmatrix} \alpha \\ \beta \end{bmatrix} = \begin{bmatrix} 0 & \beta^* \end{bmatrix} \begin{bmatrix} \alpha \\ \beta \end{bmatrix} = |\beta|^2$$

Notice that when measuring in the computational basis, the probability of getting a 0 or 1 is the magnitude squared of the probability amplitude associated with that value.

Although ideally measurement would be performed in the computational basis to read the output at the end of a quantum computation, another form of measurement often occurs. This measurement comes in the form of interference from the environment surrounding the qubits, and is known as *decoherence* or *quantum noise* [49, 61]. In practice, it is difficult to isolate stable quantum states from the environment, and since measurement of any kind is destructive, a computation can easily be ruined before it completes. This problem alone has been one of the greatest technological hurdles facing the physical realization of quantum computers [47, 59, 61].

Postulate 4. Composite quantum states are represented by the tensor product of the component quantum states, and operators that act on composite states are represented by the tensor product of their component matrices.

Simply put, this postulate enables the description of multiple qubits and multi-qubit operators via a single state vector and matrix, respectively. As defined earlier (Section 3.1), the tensor product $A \otimes B$ multiplies each element of A by the entire matrix (vector) B to produce a new matrix (vector) of higher dimension. For example,

$$A = \begin{bmatrix} a & b \\ c & d \end{bmatrix} \quad B = \begin{bmatrix} e & f \\ g & h \end{bmatrix} \quad A \otimes B = \begin{bmatrix} ae & af & be & bf \\ ag & ah & bg & bf \\ ce & cf & de & df \\ cg & ch & dg & dh \end{bmatrix}$$

Similarly, for two vectors V and W,

$$V = \begin{bmatrix} a \\ b \end{bmatrix} \quad W = \begin{bmatrix} c \\ d \end{bmatrix} \quad V \otimes W = \begin{bmatrix} ac \\ ad \\ bc \\ bd \end{bmatrix}$$

In general, there is no restriction on the dimensions of tensor product operands. Matrices of different dimensions can be tensored together, as can vectors and matrices. However, in the quantum domain, we typically perform the tensor product on square, power-of-two-sized matrices to create larger operators (Postulate 2), and also on power-of-two-sized vectors to create larger composite quantum states (Postulate 1). To illustrate, suppose we want the state vector that describes the combined state of the three separate qubits: $|1\rangle, |0\rangle, |1\rangle$. The composite state vector is computed as

$$\begin{bmatrix} 0 \\ 1 \end{bmatrix} \otimes \begin{bmatrix} 1 \\ 0 \end{bmatrix} \otimes \begin{bmatrix} 0 \\ 1 \end{bmatrix} = \begin{bmatrix} 0 \\ 0 \\ 0 \\ 0 \\ 0 \\ 1 \\ 0 \\ 0 \end{bmatrix} \tag{3.7}$$

Dirac notation offers a simple short-hand description of composite quantum states in which the state symbols are simply placed side-by-side within a single ket. For the preceding example (Equation 3.7), the Dirac form of the 3-qubit state is $|101\rangle$.

Extending the concept of measurement (Postulate 3) to composite quantum states is fairly straightforward. In the case of vectors, by multiplying each element of a vector V by an entire vector W, the tensor product produces a vector whose elements are indexed in binary counting order. To demonstrate, we revisit $V \otimes W$ annotated with binary indices assigned to the rows.

$$V \otimes W = \begin{bmatrix} a \\ b \end{bmatrix} \begin{matrix} \scriptstyle 0 \\ \scriptstyle 1 \end{matrix} \otimes \begin{bmatrix} c \\ d \end{bmatrix} \begin{matrix} \scriptstyle 0 \\ \scriptstyle 1 \end{matrix} = \begin{bmatrix} ac \\ ad \\ bc \\ bd \end{bmatrix} \begin{matrix} \scriptstyle 00 \\ \scriptstyle 01 \\ \scriptstyle 10 \\ \scriptstyle 11 \end{matrix}$$

Whereas the indices 0 and 1 for the single quantum state vectors correspond to the amount of "zeroness" and "oneness" in the quantum state, the indices in the composite vector represent the amount of "00-ness, 01-ness, 10-ness, and 11-ness," respectively. Thus, when measuring with respect to the computational basis, the binary indices of a state vector denote the classical bit values that will be measured with a probability given by the magnitude squared of the value at that location in the vector.

To illustrate the composition of quantum operators, we turn to an example involving the Hadamard operator. A Hadamard operator that can be applied to two qubits is constructed via the tensor product of two Hadamard matrices.

$$\begin{bmatrix} \frac{1}{\sqrt{2}} & \frac{1}{\sqrt{2}} \\ \frac{1}{\sqrt{2}} & -\frac{1}{\sqrt{2}} \end{bmatrix} \otimes \begin{bmatrix} \frac{1}{\sqrt{2}} & \frac{1}{\sqrt{2}} \\ \frac{1}{\sqrt{2}} & -\frac{1}{\sqrt{2}} \end{bmatrix} = \begin{bmatrix} \frac{1}{2} & \frac{1}{2} & \frac{1}{2} & \frac{1}{2} \\ \frac{1}{2} & -\frac{1}{2} & \frac{1}{2} & -\frac{1}{2} \\ \frac{1}{2} & \frac{1}{2} & -\frac{1}{2} & -\frac{1}{2} \\ \frac{1}{2} & -\frac{1}{2} & -\frac{1}{2} & \frac{1}{2} \end{bmatrix}$$

The above examples show that n qubits can be represented by $n-1$ tensor products of single-qubit vectors, and operators that act on n qubits can be represented by $n-1$ tensor products of single qubit operators. A key point to note is that the size of a state vector resulting from a series of tensor products on n single qubit vectors is 2^n. Similarly, a composite operator that can be applied to n qubits is a matrix of size $2^n \times 2^n$. It is indeed Postulate 4 which gives rise to the exponential complexity of the simulation of quantum behavior on classical computers. A straightforward linear algebraic approach to such simulation would have time and memory complexity $O(2^{2n})$ for an n-qubit system.

Quantum Properties

An interesting property of quantum states is that they cannot be arbitrarily copied, a result called the *no-cloning theorem.* [61]. This leads to yet another fundamental difference between quantum and classical computing. In classical logic circuits, a wire can fan out from the output of a gate and feed into many other gates. This copying operation is not possible in the quantum domain for an arbitrary qubit. However, it is not a major limitation because quantum states that are known to be orthogonal to each other (including the computational basis states) can be copied. A proof adapted from [61] is offered below.

Given two unknown quantum states $|\psi\rangle$ and $|\varphi\rangle$, we try to apply some unitary operator (in accordance with Postulate 2) such that both $|\psi\rangle$ and $|\varphi\rangle$ are copied to other quantum states $|s\rangle$ and $|t\rangle$. We can represent this mathematically by

$$U(|\psi\rangle \otimes |s\rangle) = |\psi\rangle \otimes |\psi\rangle$$
$$U(|\varphi\rangle \otimes |t\rangle) = |\varphi\rangle \otimes |\varphi\rangle$$

However, since quantum computing is modeled by a finite-dimensional Hilbert space, the inner product of the states defined by both equations must be defined if they are in fact valid evolutions of quantum states. The inner product of these states reduces to

$$\langle\psi|\varphi\rangle = (\langle\psi|\varphi\rangle)^2$$

Any expression of the form $x = x^2$ (as is the case above) only has two scalar solutions, $x = 0$ and $x = 1$. If the inner product of two state vectors is 0, the vectors are orthogonal. The only way for the inner product to be 1 is if both state vectors are equal. Thus, it is either the case that $|\psi\rangle$ and $|\varphi\rangle$ are orthogonal or that $|\psi\rangle = |\varphi\rangle$. This proof demonstrates that arbitrary quantum states cannot be copied. However, if it is known that the quantum states are orthogonal, they can be copied.

The implication for quantum computing is that the computational basis states $|0\rangle$ and $|1\rangle$, which are orthogonal, *can be copied.* Since these states are analogous to the classical bit values 0 and 1, the no-cloning theorem hints that quantum computers are at least as powerful as classical computers.

A standard quantum operator used to copy computational basis states (among other functions) is called the *CNOT* operator [61]. As the name implies, CNOT is a controlled-NOT operation. It is a unitary matrix (in accordance with Postulate 2) that acts on two qubits. One qubit is the *control qubit* while the other qubit is the *target qubit*. When the control qubit is in the $|1\rangle$ state, the CNOT is "activated", and the state of the target qubit is flipped from $|0\rangle$ to $|1\rangle$, or vice versa. If the control qubit is in the $|0\rangle$ state, however, the state of the target qubit is unchanged. When both the control and target qubits are in the computational basis states, the CNOT operation performs the same function as the classical XOR (exclusive-or) gate, where the target qubit receives the value of the XOR of the control qubit and the old target qubit value. To demonstrate, a CNOT operator is shown below changing the state vector $|10\rangle$ to $|11\rangle$,

$$\begin{bmatrix} 1\,0\,0\,0 \\ 0\,1\,0\,0 \\ 0\,0\,0\,1 \\ 0\,0\,1\,0 \end{bmatrix} \begin{bmatrix} 0 \\ 0 \\ 1 \\ 0 \end{bmatrix} = \begin{bmatrix} 0 \\ 0 \\ 0 \\ 1 \end{bmatrix}$$

An extension of CNOT is the *Toffoli* operator, which is basically a CNOT with two control qubits and one target qubit. In this case, the value of the target qubit is flipped if *both* of the control qubits are in state $|1\rangle$. So, given two control qubits a and b and a target qubit c, the Toffoli gate causes c to become $c \oplus ab$, where $\oplus$ denotes the XOR operator. The Toffoli gate alone is a universal gate set that can effect any form of classical circuit [61]. This is easily demonstrated by showing that the Toffoli gate can perform the same function as the NAND gate which constitutes a universal gate set. To compute a NAND b where a and b are input qubits, we simply make a and b the control qubits of the Toffoli gate and initialize the target qubit to $|1\rangle$. Such an instance of the Toffoli gate computes the function $1 \oplus ab = \neg(ab)$, which is equivalent to a NAND b. The properties of CNOTs, Hadamard gates, and other quantum gates are explored further in Chapter 4.

Another interesting property of quantum states is *entanglement*. Two quantum systems are entangled if the measurement outcome of one system affects the measurement statistics of another system, without any physical interaction at the time of measurement. A simple example of entangled states is the Bell state or EPR pair [61]. Suppose two parties, Alice and Bob, each have their own qubit, and the state of both qubits together is given as $\psi_{AB} = |0_A 0_B\rangle$, where the subscript A denotes the portion of the state due to Alice's qubit, and the subscript B denotes the portion due to Bob's qubit. An EPR pair can be generated from this state by applying a Hadamard gate and a CNOT gate as follows,

$$\psi_{EPR} = (CNOT)(H \otimes I)\,|0_A 0_B\rangle = \frac{1}{\sqrt{2}}(|0_A 0_B\rangle + |1_A 1_B\rangle)$$

The utility of this state lies in the fact that if Alice measures her particle and obtains a 0, then Bob will subsequently also obtain a 0 upon measurement of his particle (the same holds true for a measurement of 1). Once the EPR pair is created, the measurement outcomes of each qubit are correlated, even if Alice and Bob physically separate their qubits by any amount of distance. As a result, entanglement has applications in quantum teleportation [10] and secure public key exchange [9, 8, 31], which will be discussed later in Chapters 8 and 10.

Density Matrices

An important extension of the state vector is the *density matrix*. For the purposes of quantum circuit simulation, it is sufficient to define an n-qubit density matrix as $\rho = |\psi\rangle\langle\psi|$, where $|\psi\rangle$ is a single state vector for a sequence of n initialized qubits, and $\langle\psi|$ is its adjoint, i.e., its complex-conjugate transpose. Hence, ρ is a $2^n \times 2^n$ matrix constructed by multiplying a 2^n element column vector by a 2^n element row vector. For instance, when $|\psi\rangle$ is a single qubit,

$$\rho = |\psi\rangle\langle\psi| = \begin{bmatrix} \alpha \\ \beta \end{bmatrix} [\alpha^* \beta^*] = \begin{bmatrix} \alpha\alpha^* & \alpha\beta^* \\ \beta\alpha^* & \beta\beta^* \end{bmatrix}$$

As in the basic state-vector model, a gate operation U can be expressed as a density matrix, but it takes the form $U\rho U^\dagger$, where $U^\dagger$ is the complex-conjugate transpose of the matrix for U. For example, if $U = H$, the Hadamard operator,

$$H\rho H^\dagger = \begin{bmatrix} \frac{1}{\sqrt{2}} & \frac{1}{\sqrt{2}} \\ \frac{1}{\sqrt{2}} & \frac{-1}{\sqrt{2}} \end{bmatrix} \begin{bmatrix} \alpha\alpha^* & \alpha\beta^* \\ \beta\alpha^* & \beta\beta^* \end{bmatrix} \begin{bmatrix} \frac{1}{\sqrt{2}} & \frac{1}{\sqrt{2}} \\ \frac{1}{\sqrt{2}} & \frac{-1}{\sqrt{2}} \end{bmatrix}$$

Perhaps the most useful property of the density matrix is that it can accurately represent a subset of the qubits in a circuit. One can extract partial state information with the *partial trace* operation, which produces a smaller matrix, called the *reduced* density matrix [61]. To understand this extraction process, consider the following example in which a 1-qubit operator U is applied to two qubits $|\psi\rangle$ and $|\phi\rangle$. The density matrix version of this circuit is $(U \otimes U)|\psi\phi\rangle\langle\psi\phi|(U \otimes U)^\dagger = |\psi'\phi'\rangle\langle\psi'\phi'|$. The state of $|\phi\rangle$ alone after U is applied, for instance, can be extracted with the partial trace, $tr(U|\psi\rangle\langle\psi|U^\dagger)U|\phi\rangle\langle\phi|U^\dagger$. Here tr is the standard trace operation defined in Section 3.1, which produces a single complex number that is the sum of the diagonal elements of the matrix. A more concrete example is the partial trace over the first qubit in a density matrix representing two qubits with the state $\rho_0 \otimes \rho_1$, such that $\rho_0 = |+\rangle\langle+|$ and $\rho_1 = |0\rangle\langle 0|$, where $|+\rangle$ denotes an equal superposition.

$$\rho_0 \otimes \rho_1 = \begin{bmatrix} \frac{1}{2} & \frac{1}{2} \\ \frac{1}{2} & \frac{1}{2} \end{bmatrix} \otimes \begin{bmatrix} 1 & 0 \\ 0 & 0 \end{bmatrix} = \begin{bmatrix} \frac{1}{2} & 0 & \frac{1}{2} & 0 \\ 0 & 0 & 0 & 0 \\ \frac{1}{2} & 0 & \frac{1}{2} & 0 \\ 0 & 0 & 0 & 0 \end{bmatrix}$$

$$tr_{\rho_0}(\rho_0 \otimes \rho_1) = \begin{bmatrix} 1 & 0 \\ 0 & 0 \end{bmatrix}$$

Although in this example the partial trace reproduces the state of the second qubit, it does not always extract the original state of the remaining qubits. In particular, when entanglement exists among two or more qubits, the partial trace will not undo the tensor product. This issue is central to some of the simulation methods discussed in Chapter 6.

Also notice that the partial trace "traces over" the qubit that is *not wanted*, leaving behind the desired qubit states. Using the partial trace to extract information about subsets of qubits in a circuit is invaluable in simulation. As will be shown in Chapter 8, many practical quantum circuits contain ancillary qubits which help to perform an intermediate function in the circuit but contain no useful information at the output of the circuit. The partial trace therefore allows a simulation to report the density matrix information only for the qubits that contain useful data. Another application of the partial trace in quantum circuits is the modeling of noise from the environment. Coupling between the environment and data qubits can be modeled as the tensor product of data qubits with quantum states controlled by the environment [61]. In such a situation, the partial trace can be used to extract the state of data qubits after being affected by noise. For these reasons and others, it is crucial that a quantum simulator support the density matrix representation.

3.3 Summary

This chapter has reviewed both the mathematical and physical foundations of quantum information processing—linear algebra and quantum mechanics, respectively. Quantum mechanics is based on four postulates which define quantum states and operators as vectors and matrices, respectively, in a complex-valued vector space (Hilbert space). Quantum states have many strange properties such as the no-cloning property and entanglement, which have no classical counterparts and are key to the power of quantum computations. Chapter 4 builds on these principles and presents the quantum logic circuit model, which is the quantum analogue of the classical logic circuit model covered in Chapter 2.

4

Quantum Information Processing

This chapter introduces the gate and circuit models for quantum computation. To reduce the air of mystery commonly associated with quantum circuits, we compare and contrast the concepts presented here with their classical analogues discussed in Chapter 2. First, we define the basic types of quantum gates, and then show how multiple gates can be combined to create quantum circuits. In each case, we try to connect the quantum circuit concepts to their linear-algebraic underpinnings discussed in Chapter 3. This connection is critically important to quantum circuit simulation, which is explored in great detail in the remainder of the book.

4.1 Quantum Gates

Qubits can be implemented by a collection of photons, electrons, or even nuclei in an organic molecule. In these cases, information can be encoded by photon polarizations, nuclear or electronic spins. Information processing requires quantum transformations that can be applied to qubits, and these are often implemented by non-linear optical devices or pulses of electro-magnetic radiation [61]. However, most such transformations can only directly involve 1 or 2 qubits at a time, leaving other qubits unchanged. Such "small" operators are called *quantum gates* and are represented by 2×2 and 4×4 unitary matrices, respectively. While the underlying implementation technologies are often very different, a unified mathematical model has emerged for quantum gates and circuits. The physical foundations of the quantum logic gates originate in two postulates of quantum physics reviewed in Chapter 3. Postulate 1 provides the data representation, qubits, while Postulate 2 describes the action of operators on qubits. Indeed, operators are the mathematical description of the physical behavior of quantum gates, and as a result the terms operator and gate are used interchangeably.

CIRCUIT IDENTITY	DESCRIPTION
$C_j^k C_j^k = I$	CNOT-gate cancellation
$\omega^{j,k}\omega^{j,k} = I$	SWAP-gate cancellation
$C_j^k C_k^j = \omega^{j,k} C_j^k$	CNOT-gate elimination
$C_k^j R_x^j(\theta) = R_x^j(\theta) C_k^j$	moving R_x via CNOT target
$C_k^j R_z^k(\theta) = R_z^k(\theta) C_k^j$	moving R_z via CNOT control
$\sigma_x^k C_j^k = C_j^k \sigma_x^j \sigma_x^k$	moving σ_x via CNOT control
$C_j^k \sigma_z^j = \sigma_z^j \sigma_z^k C_j^k$	moving σ_z via CNOT target
$C_j^k \omega^{j,k} = \omega^{j,k} C_k^j$	moving CNOT via SWAP
$V^j \omega^{j,k} = \omega^{j,k} V^k$	moving a 1-qubit gate via SWAP
$R_n(\theta)R_n(\phi) = R_n(\theta+\phi)$	merging R_n gates.

Table 4.1. Some useful circuit identities: V^j denotes an arbitrary 1-qubit operator acting on wire j, while R_j denotes a rotation operator.

Common Gate Types

Quantum gates are usually defined by specifying their (unitary) matrices. For example, the quantum logic operation of "doing nothing" is represented by the identity matrix I_n. The quantum analogue of the classical NOT gate or inverter is specified by the following matrix:

$$\sigma_x = \begin{pmatrix} 0 & 1 \\ 1 & 0 \end{pmatrix} = |0\rangle\langle 1| + |1\rangle\langle 0|$$

which performs the mapping $|0\rangle \mapsto |1\rangle$ and $|1\rangle \mapsto |0\rangle$. It may also be noted that this is one of the three Pauli matrices introduced in Section 3.1. The other Pauli matrices define similar gates:

$$\sigma_y = \begin{pmatrix} 0 & -i \\ i & 0 \end{pmatrix} = i|1\rangle\langle 0| - i|0\rangle\langle 1|$$

and

$$\sigma_z = \begin{pmatrix} 1 & 0 \\ 0 & -1 \end{pmatrix} = |0\rangle\langle 0| - |1\rangle\langle 1|$$

Quantum gates are often specified by time-dependent matrices that represent the evolution of a quantum system (e.g., an RF pulse affecting a nuclear spin) that has been activated for some time θ. For example, the following families of 1-qubit gates are commonly available in physical implementations of quantum circuits.

- The x-axis rotation $R_x(\theta) = \begin{pmatrix} \cos\theta/2 & -i\sin\theta/2 \\ -i\sin\theta/2 & \cos\theta/2 \end{pmatrix} = \exp(pi\theta\sigma_x)$
- The y-axis rotation $R_y(\theta) = \begin{pmatrix} \cos\theta/2 & -\sin\theta/2 \\ \sin\theta/2 & \cos\theta/2 \end{pmatrix} = \exp(-i\theta\sigma_y)$

- The z-axis rotation $R_z(\theta) = \begin{pmatrix} e^{-i\theta/2} & 0 \\ 0 & e^{i\theta/2} \end{pmatrix} = \exp(-i\theta\sigma_z)$

An arbitrary 1-qubit computation can be implemented as a sequence of at most three R_z and R_y gates. This is due to the well-known *ZYZ decomposition* which states: given any 2×2 unitary matrix U, there exist angles $\Phi, \theta_0, \theta_1, \theta_2$ satisfying the following equation.

$$U = e^{i\Phi} R_z(\theta_0) R_y(\theta_1) R_z(\theta_2)$$

Similar results hold for the ZXZ, YZY, YXY, XZX and XYX configurations.

The nomenclature R_x, R_y, R_z is motivated by a picture of 1-qubit states as points on the surface of a sphere of unit radius in three-dimensional space, called the *Bloch sphere* [61]. It is obtained by expanding an arbitrary two-dimensional complex vector as shown below.

$$|\psi\rangle \;=\; \alpha_0 |0\rangle + \alpha_1 |1\rangle \;=$$
$$re^{it/2}\left[\; e^{-i\varphi/2} \cos\frac{\theta}{2} |0\rangle + e^{i\varphi/2} \sin\frac{\theta}{2} |1\rangle \;\right]$$

The constant factor $re^{it/2}$ is physically undetectable. On ignoring it, we are left with two angle parameters θ and φ, which we interpret as spherical coordinates $(1, \theta, \varphi)$. In this picture, $|0\rangle$ and $|1\rangle$ correspond to the north and south poles, $(1, 0, 0)$ and $(1, \pi, 0)$, respectively. The $R_x(\theta)$ gate corresponds to a counterclockwise rotation by θ around the x axis; $R_y(\theta)$ and $R_z(\theta)$ perform similar rotations around the y and z axes, respectively. Finally, just as the point given by the spherical coordinates $(1, \theta, \varphi)$ can be moved to the north pole by first rotating $-\varphi$ degrees around the z-axis, then $-\theta$ degrees around the y axis, so too the following matrix equations hold.

$$R_y(-\theta) R_z(-\varphi) |\psi\rangle = re^{it/2} |0\rangle$$
$$R_y(\theta - \pi) R_z(\pi - \varphi) |\psi\rangle = re^{i(t-\pi)/2} |1\rangle$$

One of the simplest 2-qubit gates is the *controlled-NOT* (CNOT) gate mentioned in Section 3.2. It has the following matrix representation and graphic symbol:

$$\text{CNOT} = \begin{bmatrix} 1\,0\,0\,0 \\ 0\,1\,0\,0 \\ 0\,0\,0\,1 \\ 0\,0\,1\,0 \end{bmatrix} = \tag{4.1}$$

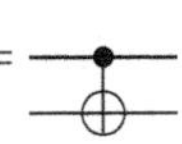

The CNOT gate acts on two qubits: it flips the second (target) qubit b_2 if the first (control) qubit b_1 is $|1\rangle$, hence the name "controlled-NOT." CNOT has its own graphic symbol for use in circuit diagrams: a "$\bullet$" on the control qubit connected by a vertical line to an "$\oplus$" on the target qubit. This notation is motivated by the formula $|b_1\rangle |b_2\rangle \mapsto |b_1\rangle |b_1 \text{ XOR } b_2\rangle$, which relates CNOT to

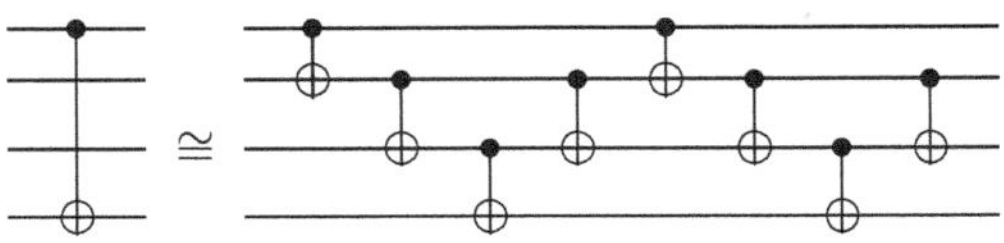

Fig. 4.1. Implementing a long-range CNOT gate with adjacent-qubit CNOTs.

the classical XOR gate. Several CNOTs are illustrated in the quantum circuits of Figure 4.1.

In 2-qubit circuits, we sometimes distinguish between $\text{CNOT}_{top} = |00\rangle\langle 00| + |01\rangle\langle 01| + |10\rangle\langle 11| + |11\rangle\langle 10|$ which describes the CNOT gate with control on the upper qubit (shown in Equation 4.1 above), and its upside-down variant $\text{CNOT}_{bot} = |00\rangle\langle 00| + |10\rangle\langle 10| + |01\rangle\langle 11| + |11\rangle\langle 01|$ with control on the lower qubit. It is easy to check that $\text{CNOT}_{top}|11\rangle = |10\rangle$, while $\text{CNOT}_{top}|11\rangle = |01\rangle$.

The CNOT gate, together with the 1-qubit gates defined above form a gate library for quantum circuits that is *universal* in the sense that any unitary operator can be decomposed into CNOTs and those 1-qubit gates. Given that CNOT gates are typically the most expensive, take the longest time, and are more susceptible to errors, Shende, Bullock and Markov have shown that worst-case n-qubit operators require between $\lceil \frac{1}{4}(4^n - 3n - 1) \rceil$ and $(23/48) \times 4^n - (3/2) \times 2^n + 4/3$ CNOT gates [75, 76]. In particular, any 2-qubit operator can be decomposed into 1-qubit gates and up to three CNOT gates [75].

Among other notable 1-qubit gates, we highlight the *Hadamard (H) gate*

$$H = \frac{1}{\sqrt{2}} \begin{pmatrix} 1 & 1 \\ 1 & -1 \end{pmatrix}$$

which was also introduced in Chapter 3. This gate maps $|0\rangle$ to $(|0\rangle + |1\rangle)/\sqrt{2}$ and can be viewed as a coin toss—starting from a known state (heads), a superposition of heads and tails is created. When this superposition is measured, the coin assumes the heads or tails state with equal probability.

The *phase gate* is defined as

$$P(\theta) = \begin{pmatrix} 1 & 0 \\ 0 & e^{2\pi i\theta} \end{pmatrix}$$

It is most often used with $\theta = \pi/8$, in which case it is called the $\pi/8$-gate and denoted by P. The H, P and CNOT gates are collectively called *stabilizer gates*, and are the building blocks for most circuits used in quantum error correction.

As observed earlier, the three-input Toffoli gate T is an extended version of the CNOT gate, which inverts a target qubit when its two control qubits carry 1. Its 8×8 matrix can be compactly written in block-diagonal form as $1 \oplus 1 \oplus \sigma_x$. One draws a Toffoli gate just like the CNOT gate, but with an additional control wire.

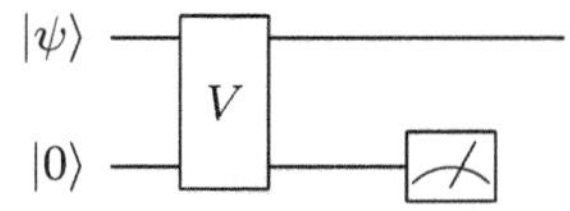

Fig. 4.2. General gate model for environmental noise on a qubit ($|\psi\rangle$.

The idea of adding control qubits to any quantum gate U gives rise to the class of *controlled-U* or $C(U)$ gates defined by $1 \oplus U$. In other words, when the control qubit carries $|1\rangle$, the U operation is applied to the target qubit(s). When the control qubit carries $|0\rangle$, U is not applied. The behavior on other inputs is defined by linear interpolation from these basis cases.

Density-matrix Formalism for Gates

Section 3.2 mentions using density matrices and partial traces to compute the effect of environmental noise on a single qubit $|\psi\rangle$. A gate-level schematic that models this scenario is shown in Figure 4.2. Typically, the second qubit, which represents the state of the environment, is initialized to the ground state, $|0\rangle$.

Different 2-qubit gates may be substituted for V to induce different levels and types of coupling between the data qubit state $|\psi\rangle$ and the environment [61]. To connect the gate-level schematic to the density matrix model, recall that the first step is to create the density matrix for the initial state via the outer product operation,

$$\rho_{in} = |\psi\ 0\rangle\ \langle\psi\ 0|$$

The resulting operator is then applied to the qubits by premultiplying the density matrix by V and postmultiplying it by the adjoint operator $V^\dagger$:

$$\rho_{out} = V \rho_{in} V^\dagger$$

Notice that both ρ_{in} and ρ_{out} represent 2-qubit states. This point is particularly important in the case of ρ_{out} which, in general, may no longer be decomposable into a single tensor product of distinct 1-qubit density matrices due to the entangling action of V (such states are called *inseparable.*) The creation of entanglement in circuits is described in Section 4.2, while advanced techniques for decomposing entangled states into sums of tensor products are described in Chapter 6. Before moving on to those topics, it is instructive to finish analyzing the gate-level schematic in Figure 4.2.

The meter symbol on the lower qubit denotes a measurement operation performed on the environment. Recall that quantum measurement is destructive and produces a deterministic outcome (Postulate 3). Therefore, the effect of coupling $|\psi\rangle$ to the environment through V followed by measurement on the environment qubit induces a kind of partial deterministic measurement on the data qubit, forcing it to be closer to a classical state. To extract the altered state of the data qubit, the partial trace may be used after measurement to trace over the environment's contribution to the 2-qubit state, leaving only the state of the data qubit.

There are other situations in which tracing over some subset of the qubits to examine the state of the remaining qubits is useful. Generally speaking, all these cases involve observing the effects of entanglement in various types of circuits. The following section explains how multiple gates and qubits fit together, explores the role of entanglement, and investigates how different gates can generate entanglement.

4.2 Quantum Circuits

Since the first successful demonstrations of quantum gates and circuits by applied physicists in the 1990s, the quantum logic circuit model was proven itself useful in nanoscale systems with a wide variety of unrelated physical characteristics. This level of abstraction is facilitated by the postulates of quantum mechanics reviewed in Chapter 3. The two major components of any quantum circuit are the qubits (Postulate 1) and the operators or gates (Postulate 2). The values of the qubits are observed through measurement (Postulate 3), and multiple qubits and gates can be expressed via the tensor product (Postulate 4). The postulates of quantum mechanics provide a complete set of properties with which to perform quantum logic design.

Taking all the foregoing properties into account, a *quantum circuit* is defined by:

- An ordered set of qubits
- A sequence of gate operations applied to the qubits

Each gate G in the circuit can be specified by a small matrix or its graphic symbol, the indices of the qubits to which G is applied, and G's position in the sequence of gate operations. In the remainder of this section, we present two small examples to familiarize the reader with the standard quantum circuit notation.

Reversible Half-adder

The first example, shown in Figure 4.3, is a quantum half-adder that computes the sum and carry functions of two qubits $|A\rangle$ and $|B\rangle$. It performs the same functions as the standard half-adder of classical logic when the inputs are confined to the computational basis states $|0\rangle$ and $|1\rangle$. A third qubit C is also involved in the quantum half-adder, whose role will be explained shortly. Notice that the qubits are depicted graphically as parallel, horizontal lines.

CNOT * Toffoli * $(|A\rangle \otimes |B\rangle \otimes |C\rangle)$

(a)

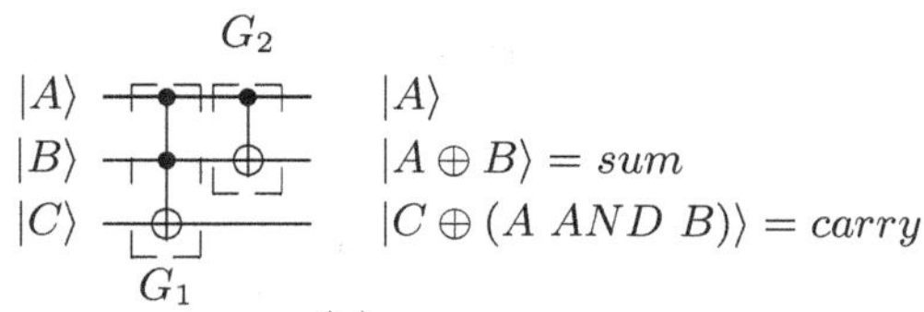

(b)

Fig. 4.3. (a) Textual and (b) graphical depiction of a reversible quantum half-adder circuit.

These lines can be thought of as wires, but it is more realistic to see them as representing the evolution of the qubits over time, where time increases from left to right. Gates are depicted by symbols placed on top of the horizontal qubit lines; they affect only those qubit lines that they touch. The spacing between the gates on the qubit lines has no significance. The key aspect of gate placement is that it defines the order in which the gate operations are applied to the affected qubits. In general, the input states of the qubits are placed at the left end of the qubit lines, with the final output states of the qubits appearing at the right end. Figure 4.3b is therefore intended to suggest that the three input states $|A\rangle$, $|B\rangle$ and $|C\rangle$, are transformed (evolve) over time into the output states $|A\rangle$, $|sum\rangle$ and $|carry\rangle$.

The quantum half-adder contains just two gates: a Toffoli gate G_1 affecting all three qubits, followed by a $CNOT$ gate G_2 affecting the first two qubits only. The solid circles mark the control inputs of the qubits, while the $\oplus$ symbols mark the target qubits. The matrix representation of the half-adder circuit is

$$\text{H\_Adder} = (G_2 \otimes I)G_1 = \left(\begin{bmatrix} 1&0&0&0 \\ 0&1&0&0 \\ 0&0&0&1 \\ 0&0&1&0 \end{bmatrix} \otimes \begin{bmatrix} 1&0 \\ 0&1 \end{bmatrix} \right) \begin{bmatrix} 1&0&0&0&0&0&0&0 \\ 0&1&0&0&0&0&0&0 \\ 0&0&1&0&0&0&0&0 \\ 0&0&0&1&0&0&0&0 \\ 0&0&0&0&1&0&0&0 \\ 0&0&0&0&0&1&0&0 \\ 0&0&0&0&0&0&0&1 \\ 0&0&0&0&0&0&1&0 \end{bmatrix} \tag{4.2}$$

Several important aspects of quantum computation can be deduced from the above equation. The combined effect of the Toffoli gate G_1 followed by the CNOT gate G_2 is represented by the ordinary matrix product. Although the circuit action flows from left to right, the matrices representing the operators are written in the seemingly reverse right-to-left order; this is merely a consequence of the normal mechanics of matrix multiplication [89].

The action of the Toffoli gate on the half-adder's three qubits is represented by its defining 8×8 matrix $G_1 = T$. To represent the action of the CNOT

$$(H \otimes H)(|A\rangle \otimes |B\rangle)$$

(a)

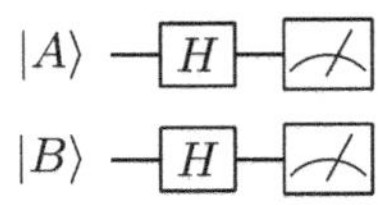

(b)

Fig. 4.4. (a) Textual and (b) graphical depiction of a quantum circuit which places two qubits into an equal superposition when $|A\rangle$ and $|B\rangle$ are initialized to $|0\rangle$.

on all three qubits, we must combine the CNOT's 4×4 matrix G_2 with the 2×2 identity (null) matrix I acting on the bottom qubit. This combination is represented by the tensor product $G_2 \otimes I$, as indicated in Equation 4.2, which evaluates to an 8×8 matrix. The final 8×8 matrix product $(G_2 \otimes I)G_1$ is therefore the matrix specification for a quantum half-adder.

Several other features of Figure 4.3b distinguish it from a classical logic circuit. There is no branching or fanout of wires, a consequence of the no-cloning theorem (see Section 3.2). There is an extra *ancilla* qubit $|C\rangle$, apparently giving rise to some superfluous input and output lines. However, ancilla qubits are a necessary consequence of the unitarity or reversibility of quantum operations, which requires a quantum circuit to carry enough extra information to allow the original input state of the qubits to be reconstructed from knowledge of its output state. The ancillae are sometimes described as "scratchpad" variables that record data about past states. Finally, although it is often convenient to think of Figure 4.3b as a combinational circuit, it actually describes a sequence of computational steps, and so resembles a classical sequential circuit.

Creating Some Useful Quantum States

Figure 4.4 illustrates creation of the balanced superposition $(|0\rangle + |1\rangle)/\sqrt{2}$ on each of two qubits using a pair of Hadamard gates. As a result, the two qubits together carry the balanced 2-qubit superposition $(|00\rangle + |01\rangle + |10\rangle + |11\rangle)/2$. Layers of n H gates applied to sets of n qubits in this way are often used in quantum algorithms to allow subsequent gates to operate on every input combination at once—this is the basic form of quantum parallelism. Such parallel computations change the amplitudes of individual qubits, but the 2^n new amplitudes cannot be measured directly. The circuit of Figure 4.4b illustrates the measurement of individual qubits whose probabilistic outcome is a direct reflection of their amplitudes. Unfortunately, to repeat the measurement it is necessary to rerun the entire sequence of gate operations specified by the quantum circuit.

An important property of quantum states, which was introduced in Section 3.2, is entanglement. The quantum circuit of Figure 4.5, which consists of a Hadamard gate followed by a CNOT, is designed to produce the highly

$\mathrm{CNOT}_{top}(H \otimes I) =$ 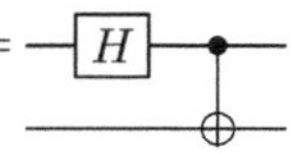

Fig. 4.5. A quantum circuit that generates the entangled EPR state.

entangled Bell or EPR-pair state $\frac{|00\rangle+|11\rangle}{\sqrt{2}}$. On applying the circuit to the 2-qubit state $|00\rangle$, the Hadamard gate H produces the state $\frac{|0\rangle+|1\rangle}{\sqrt{2}}|0\rangle$, which can be rewritten as $\frac{|00\rangle+|10\rangle}{\sqrt{2}}$ This state is then mapped by the CNOT gate to $\frac{|00\rangle+|11\rangle}{\sqrt{2}}$.

Gate Libraries

Universal quantum gate libraries can implement any quantum computation, but several interpretations of this concept exist. *Discrete* gate libraries contain only a finite number of gates, and thus cannot include parametrized families, such as R_x, R_y or R_z gates. Circuits of gates from a discrete library can only approximate arbitrary quantum computations, though the approximation can achieve any desired level of accuracy [61]. An example of a discrete universal quantum gate library consists of the Hadamard, phase, CNOT and $\pi/8$ gates. Another four-member universal library is given by the Hadamard, phase, CNOT and Toffoli gates.

Universal quantum gate libraries containing parametrized gate families allow the exact implementation of any quantum computation. One such library consists of the CNOT gate and the infinite set of all 1-qubit unitary operators [61]. Other universal libraries can be formed by taking the CNOT gate and any two members of the R_x, R_y or R_z families. For reversible circuits that implement only classical computations, the NOT, CNOT and Toffoli gates suffice, since they are functionally equivalent to the classical universal gate libraries consisting only of the NAND gate. One reason why NAND is only universal for classical computation is that it does not create entanglement. In contrast, the stabilizer gate library consisting of the CNOT, Hadamard and phase gates can create entanglement, but it is not universal [1, 36]. The significance of the stabilizer gate library will be discussed in Chapter 5.

Interestingly, it has been shown that the CNOT gate requires only the addition of a single 1-qubit gate of a specific kind to become a universal gate library for quantum computation [80]. The efficiency of a universal gate library can be estimated by the number of gates required to implement an arbitrary quantum operation, especially the number of CNOTs required, since these gates are often slower and more error-prone than 1-qubit gates. To this end, the Solovay-Kitaev theorem [48, 61] states that an approximation with desired accuracy ϵ can be achieved using discrete gate libraries with only poly-logarithmically more gates as a function of ϵ as compared to fully universal gate libraries. Therefore, discrete universal gate sets may be of practical value, especially that they are more amenable to quantum error-correction than parametrized gate families.

4.3 Synchronization of Quantum Circuits

While quantum circuit diagrams look similar to the diagrams used for classical digital circuits, the differences in implementation dramatically affect the notions of circuit timing, synchronization and sequential computation, which were reviewed in Section 2.3.

Classical digital circuits, such as those implemented in silicon chips, consist of physical gates interconnected by wires. Logical signals are implemented by electrical signals that travel through the gates and wires at a speed limited by the speed of light (roughly 300,000 km/s). The travel time determines the maximum frequency or, equivalently, the minimum clock period, at which the circuit can be reliably operated. The same holds true for quantum circuits implemented in quantum-optical technologies, where photons represent qubits that "fly" through optically active media. While optics has been used in quantum *communication* with great success, practical quantum-optical *computation* has only been demonstrated recently [54], partly because photons cannot be stopped and synchronized easily [61].

The leading particle-based quantum computing technologies, including nuclear magnetic resonance and trapped ions, employ stationary rather than photon-style flying qubits. They apply quantum gates to particle qubits by means of laser or RF pulses. The timing of each pulse and its waveform are controlled by a classical computer, which schedules the entire pulse train needed to implement a quantum computation. This arrangement makes the timing analysis of quantum circuits independent of quantum physics and relegates clocking issues to the classical computer that drives the quantum circuit, while possibly allowing the parallel application of quantum gates on non-overlapping qubits. More importantly, quantum circuits that operate on stationary qubits behave like classical sequential circuits in which all combinational gates are separated by flip-flops or registers. Unlike most classical circuits, however, quantum circuits can pause after a gate operation is applied and resume later on. No "quantum flip-flops" are required to sample the quantum state, since this state is inherently stationary rather than transient.

While the traditional notion of a synchronous sequential circuit does not lend itself particularly well to quantum computation, the finite-state machine concept has a direct and useful analogue in a quantum finite automaton (QFA) [60, 15]. The state space of a QFA is a finite-dimensional vector space on which state transitions act in the form of unitary operators representable by quantum circuits. The selection of a particular operator (a gate or circuit) is controlled by (non-quantum) configuration bits, either through the classical computer driving the QFA, or by means of controlled gates of the kind described earlier, such as CNOT and Toffoli gates.

A QFA can model several quantum circuits applied to the same qubits simultaneously, or a single circuit that is iterated many times, e.g., the main loop of Grover's algorithm discussed in the next section. An additional aspect of QFAs is that they can apply a quantum measurement to some or all of their qubits at the end of a state transition. In many quantum algorithms, such a measurement is applied at the very end in order to convert quantum information into a non-quantum final result. As long as the runtime of each QFA state transition (step) does not depend on the number of qubits and remains constant (possibly accounting for parallelism), we can compare the runtime of quantum algorithms by counting steps rather than gates.

4.4 Sample Algorithms

Quantum algorithms can be viewed as quantum circuits designed to carry out particular computations. Consequently, the terms quantum circuit and quantum algorithm are often used interchangeably in the literature, and this book is no exception. In this section, the high-level functionality of two well-known quantum algorithms is discussed. First, we describe unstructured quantum search by means of Grover's algorithm. Shor's algorithm for polynomial time integer factorization is then covered. QuIDDPro code for simulating both algorithms is found in Appendix B.

Unstructured Quantum Search

Grover's algorithm searches for a subset of desired items in an unordered database of N items. The only selection criterion available is a black-box predicate or oracle that can be evaluated on any item in the database. The complexity of this evaluation (query) is unknown, and the overall complexity analysis is performed in terms of queries. In the classical domain, any algorithm for such an unordered search must query the predicate $\Omega(N)$ times. However, Grover's algorithm can perform the search with quantum query complexity $O(\sqrt{N})$, a quadratic improvement in speed. This assumes that a quantum version of the search predicate can be evaluated on a superposition of all database items.

A sample circuit implementing Grover's algorithm is shown in Figure 4.6. (See Appendix B for a fully functional QuIDDPro implementation of the algorithm.) It involves five major components:

- An oracle
- A conditional phase-shift operator
- Several sets of Hadamard gates
- The data qubits

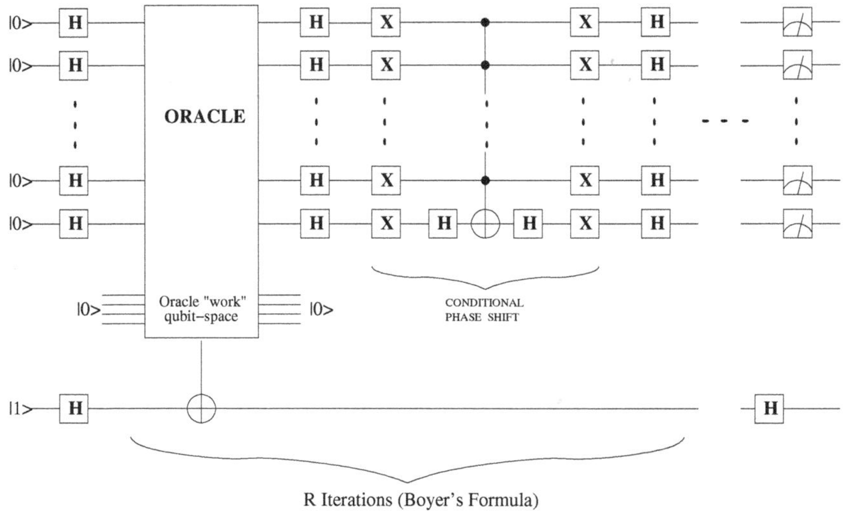

Fig. 4.6. Circuit implementation of Grover's unstructured search algorithm. Here H and X denote Hadamard and NOT gates, respectively.

- An oracle qubit

The oracle is implemented as a Boolean predicate that acts as a filter, flipping the oracle qubit when it receives as input an n-bit sequence representing the items being searched for. In quantum circuit form, the oracle can be represented by a series of CNOT gates with subsets of the data qubits acting as the control qubits, and the oracle qubit receiving the action of the CNOTs. Following the oracle circuit, Hadamard gates put the n data qubits into an equal superposition of all 2^n items in the database, where $2^n = N$. Then a sequence of gates $H^{\otimes n-1}PH^{\otimes n-1}$, where P denotes the phase-shift operator, are applied iteratively to the data qubits. Each such iteration is termed a *Grover iteration* [61].

Grover's algorithm must be stopped after a specific number of iterations when the probability amplitudes of the states representing the items sought are sufficiently large. There must be enough iterations to ensure a successful measurement of the result. After a certain point, the probability of a successful measurement starts to decline, and later it fluctuates periodically. In our simulation experiments with QuIDDPro, we used the tight bound on the number of iterations formulated by Boyer et al. [17] when the number of solutions M is known in advance: $\lfloor \pi/4\theta \rfloor$ where $\theta = \sqrt{M/N}$. The power of Grover's algorithm lies in the fact that the data qubits store all $N = 2^n$ items in the database as a superposition of quantum states, allowing the oracle circuit to find all items being searched for *simultaneously*.

Quantum Integer Factorization

The best-known classical algorithms for finding non-trivial factors of integers require runtime that is sub-exponential but super-polynomial in the number of bits of the integers. In contrast, Shor's quantum integer factorization algorithm achieves polynomial runtime complexity [82]. Descriptions of this algorithm can be quite complicated when delving into the details of phase estimation and number theory, each of which is heavily used by the algorithm. However, an intuitive understanding of how quantum mechanics is exploited to speed up factoring can be readily developed, as the following example shows.

Suppose we wish to factor the integer 15. We will perform some seemingly peculiar steps to do this, and later explain what the steps mean in general. First, select a value between 1 and 15, say 7. Now calculate successive powers of 7 modulo 15: 7^1 *mod* $15 = 7$, 7^2 *mod* $15 = 4$, 7^3 *mod* $15 = 13$, 7^4 *mod* $15 = 1$, 7^5 *mod* $15 = 7$, 7^6 *mod* $15 = 4$, 7^7 *mod* $15 = 13$, 7^8 *mod* $15 = 1$, ... Clearly there is a pattern to the results as they repeat every four powers of 7. In number-theoretic terms, the order r of 7 *mod* 15 is 4. Finally, compute $7^{r/2} - 1$, and use the result, 48, as an argument to the classical greatest common divisor algorithm along with the integer we wish to factor, 15. The greatest common divisor of 48 and 15 is 3, which is one of the factors of 15.

In general, the goal of this procedure is to compute $f(x) = a^x \; mod \; N$, where N is the integer to be factored, a is a randomly chosen value for the range $(1..N)$, and x is set to successive powers up to N. For the first (smallest) x such that $f(x) = 1$, the value of x is the order. Skipping the number-theoretic details, once the order is obtained it is known that one or both non-trivial factors of N may be obtained by computing $gcd(a^{r/2} - 1, N)$ and $gcd(a^{r/2} + 1, N)$. In the example above, we computed the first expression to obtain 3. Alternatively, by evaluating the second expression, $gcd(7^2 + 1, 15)$, we obtain a second non-trivial factor of 15, namely 5. This second calculation is academic since verification of the first factor by dividing N evenly reveals a second non-trivial factor of N.

The runtime complexity of the foregoing procedure for a given N is exponential in the number of bits required to represent N, since, in the worst case. we must try all values of $f(x)$ with x ranging from 1 to N. This is precisely where Shor's algorithm comes into the picture. Through a clever combination of Hadamard and classical reversible gate operations, Shor's algorithm computes *a superposition of outcomes for* $f(x)$ *entangled with a superposition of all possible inputs* x. In other words, Shor uses entanglement and superposition to compute all possible outcomes for $f(x)$ *simultaneously* [82].

The circuit implementation details of Shor's algorithm can be quite complicated. However, it is enough to note the key steps in the circuit depicted in Figure 4.7. First, two registers of qubits are initialized. One register will contain the superposition for the inputs x, and the other will contain a superposition of the outputs of $f(x)$. The number of qubits n in both registers is on the order of $\log N$. Different implementations use various multiples of this

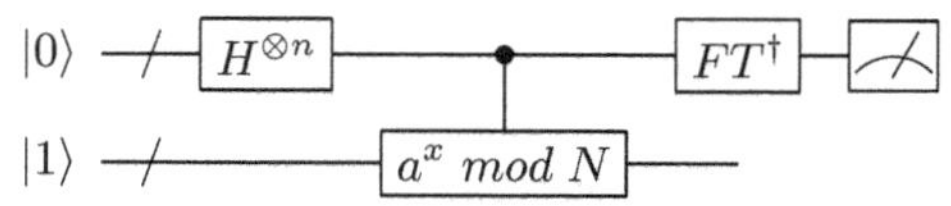

Fig. 4.7. Circuit implementation of Shor's factoring algorithm.

value to increase accuracy and make trade-offs between runtime, memory (in qubits), and circuit topology constraints such as nearest-neighbor restrictions on communication [93, 92, 33].

As shown in the figure, the second register is typically initialized to the state $|1\rangle$, since the modular exponentiation sub-circuit typically multiplies many products of a in sequence using classical reversible gates (often many adders are cascaded in sequence [93]). After the superposition on x is created, the modular exponentiation sub-circuit generates a superposition of $f(x)$ outputs. At this point, the registers have the following state: $\sum_1^N |x\rangle |a^x \; mod \; N\rangle$. Now, every non-zero state in the superposition has equal probability of being measured. Therefore, the amplitudes of those states with $|a^x \; mod \; N\rangle = |1\rangle$ must be amplified. This final step is accomplished by the inverse Quantum Fourier Transform. With the order measured from the first register with high probability, classical post-processing with the *gcd* algorithm may be performed as described above to find the factors. If the final result is not verified to be a non-trivial factor of N, the procedure is repeated with a new value for a. Appendix B.3 exhibits a QuIDDPro implementation of this algorithm.

A robust implementation of Shor's algorithm would also include classical pre-processing to weed out any integer N that is trivial to factorize. For example, any even integer always has the non-trivial factor 2, and any integer that ends in 5 has a non-trivial factor of 5. In addition to such checks, it may be worth performing the polynomial time primality test to weed out prime numbers which only have trivial factors of 1 and N.

4.5 Summary

This chapter has presented the main features of the quantum circuit model and their underlying linear-algebraic representation. Later chapters leverage this circuit-algebra connection to develop efficient algorithms for quantum circuit simulation. The key idea is often to exploit various special matrix structures to accelerate standard algorithms for inner products, matrix multiplication, tensor products, etc. The next three chapters introduce specific examples of such techniques, starting with the stabilizer formalism in Chapter 5. Chapter 6 expands this discussion with detailed descriptions of a variety of other techniques that exploit a wide range of interesting linear algebraic properties. Later, Chapter 7 will introduce the QuIDD data structure, which systematically compresses matrices and operates on the compressed forms directly using the BDD techniques discussed in Chapter 2.

5

Special Case: Simulating Stabilizer Circuits

We begin the discussion of quantum circuit simulation techniques with an in-depth analysis of a well-known technique which makes use of group theory to efficiently simulate a particular class of quantum circuits. The background material from Chapters 2, 3 and 4 will be combined to show how this class of quantum circuits, namely stabilizer circuits, and a corresponding compact representation, the stabilizer formalism, make up some key elements of a quantum circuit simulator. This special case study will set the stage for understanding the overall structure of a variety of other quantum circuit simulation techniques covered in the remainder of the book.

5.1 Basics of a Quantum Circuit Simulator

The operations defined for quantum circuits in Chapter 4 describe in abstract terms the evolution of a quantum state vector. If the circuits are implemented physically in hardware, the output is likely to be the result of a state measurement, since the postulates of quantum mechanics dictate that a state is measured when it is observed (see Chapter 3). If, however, the circuit is implemented by simulation on a classical computer, the output states or operators may be observed mathematically prior to measurement. A quantum circuit simulator can thus be defined as an abstract interface to quantum evolution, including a circuit description and its input state. Such a simulator must provide at least the information produced by an ideal quantum computer (i.e., the results of measurements), and possibly more (mathematical descriptions of states before measurements).

As indicated at the outset, we will address a number of classical computer implementations of quantum circuit simulators. To this end, the inputs and operators which make up the quantum circuit must be represented in the host computer's memory. However, as shown in Chapter 3, explicit vector and matrix representations have a size that grows exponentially with the number of qubits involved. As a result, classical simulation is feasible only when the inputs and operators or the target circuit can be represented in a compact or compressed way.

A recurring theme in the classical simulation of quantum circuits is the design of techniques which exploit the special properties of restricted classes of states, gates and/or circuit topologies. As will be shown shortly, the members of such classes often exhibit some sort of unintentional redundancy, which can be eliminated while preserving all of the pertinent information about the circuit being simulated.

Compressing redundancy in the linear-algebraic representation of a state, gate, or circuit topology requires specialized data structures. Such data structures may make use of graphs, hash tables, and any number of other representations from computer science and mathematics. Compared with the typical array implementation of explicit vectors and matrices, the combination of these data structures can be quite a bit more advanced, with various specialized rules for maintaining and updating the underlying structure.

Complementing the data structures are algorithms that manipulate the data structures directly without decompressing them to extract hidden redundancy. Direct manipulation is critical since temporarily converting the data structure to some less efficient representation, such as an array containing explicit vectors or matrices, in order to apply a gate or perform measurement does not improve the overall asymptotic runtime and memory performance. As a result, these classical algorithms can be as specialized as the data structures they manipulate.

In order to produce a quantum circuit output, classical algorithms need to simulate several types of actions that can occur. These actions include input preparation for n-qubit states and operators, the application of n-qubit quantum gates to states and other operators, and qubit measurement. Quantum circuit simulators may incorporate other helper algorithms for convenience, but at a minimum these three actions must have suitable algorithmic implementations, even when the class of quantum circuits being simulated is restricted. Figure 5.1 depicts the abstract interface of a quantum circuit simulator and indicates how the various components may be implemented.

To illustrate the requirements of quantum circuit simulation, recall the description of Shor's algorithm for integer factorization [82] discussed in Section 4.4. When computing multiple inputs of the function $f(x) = a^x \mod N$ to factor N, data structures representing the registers containing the inputs x and the outputs $f(x)$ must be initialized either to the ground state or to a superposition of values in the case of x. Additionally, the action of the modular exponentiation circuit and inverse quantum Fourier transform must be applied algorithmically to these data structures. The final output, namely the order of a^x *mod* N must be extracted from the data structures via one or more algorithms that simulate measurement.

To summarize, a practical quantum circuit simulator implemented as a program on a classical computer relies on the following key concepts.

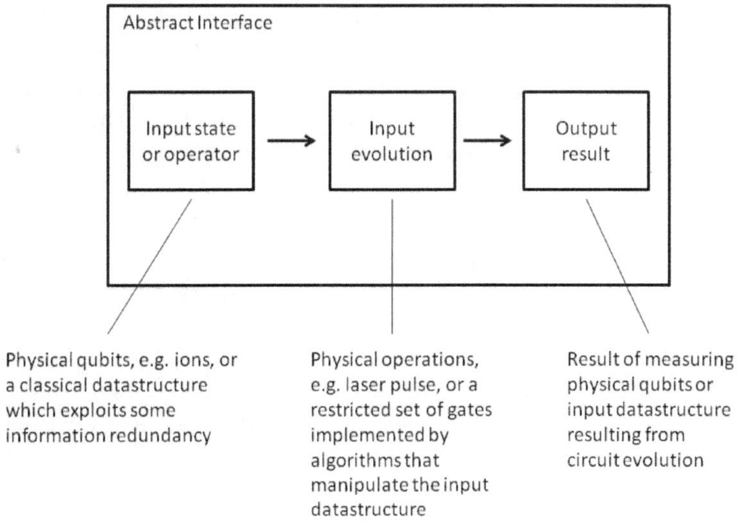

Fig. 5.1. The main components of a quantum circuit simulator and their implementation.

- A restricted domain of states, gates, and/or circuit topologies which exhibit some form of information redundancy in their mathematical representation.
- Classical data structures which can compress any redundancy present in the restricted domain.
- Classical algorithms that act directly on these data structure and efficiently implement input state and operator preparation, gate application, and qubit measurement.

5.2 Stabilizer States, Gates and Circuits

A well-known class of quantum circuits, which is essential for quantum error-correction, is distinguished by the so-called stabilizer formalism [36]. The key concept here is the *Heisenberg representation* of quantum mechanics, which tracks the commutators of a particular group of operators applied in a quantum circuit [36]. (The commutator of two operators F and G is defined as $[F, G] = F^{-1}G^{-1}FG$.) With this approach, the state need not be represented explicitly by a state vector or a density matrix because the operators describe how an arbitrary state is altered by the circuit.

The term *stabilizer* is generally applied to operators that, when applied to all qubits in a state vector, do not alter the vector. To illustrate, consider the 2-qubit computational basis state $|00\rangle$. Also recall the Pauli matrices introduced in Section 3.1. We now give them slightly different labels which are more convenient for the stabilizer formalism, as well as a graphical representation in the form of 1-qubit gates; see Figure 5.2. It is also convenient to add the 1-qubit identity matrix I to the Pauli matrices, thereby creating a four-member set of operators which forms an algebraic group with respect to matrix multiplication.

$$I = \begin{bmatrix} 1 & 0 \\ 0 & 1 \end{bmatrix} \; X = \sigma_x = \begin{bmatrix} 0 & 1 \\ 1 & 0 \end{bmatrix} \; Y = \sigma_y = \begin{bmatrix} 0 & -i \\ i & 0 \end{bmatrix} \; Z = \sigma_z = \begin{bmatrix} 1 & 0 \\ 0 & -1 \end{bmatrix} \quad (5.1)$$

Notice that $(Z \otimes I)|00\rangle = (I \otimes Z)|00\rangle = |00\rangle$. In other words, neither of the operators $(Z \otimes I)$ nor $(I \otimes Z)$ alters the state $|00\rangle$. As a result, the two operators $(Z \otimes I)$ and $(I \otimes Z)$ may be used interchangeably with the state $|00\rangle$, making these operators stabilizers. Thus, if we apply a Hadamard gate to both qubits of the $|00\rangle$ state, as in $(H \otimes H)|00\rangle$, we may alternatively compute $H(Z \otimes I)H^\dagger$ and $H(I \otimes Z)H^\dagger$.

In general, an arbitrary n-qubit state can be represented by $O(2^n)$ stabilizers [1]. At first glance, this does not seem an improvement over explicit simulation with arrays of state vectors and operator matrices. However, something interesting arises when carrying out the alternative computation. It turns out that $H(Z \otimes I)H^\dagger) = (X \otimes I)$ and $H(I \otimes Z)H^\dagger = (I \otimes X)$ Notice that applying the Hadamard operator to these particular input stabilizers produces two new stabilizers whose component matrices are Pauli matrices, just like the input stabilizers. The Pauli matrices are actually members of the Clifford group [36]. In addition to the Pauli matrices, the matrices representing the $CNOT$, phase (S), and Hadamard operators (introduced in Chapter 4) are also members of the Clifford group. The Clifford group is the *normalizer* of the group of Pauli

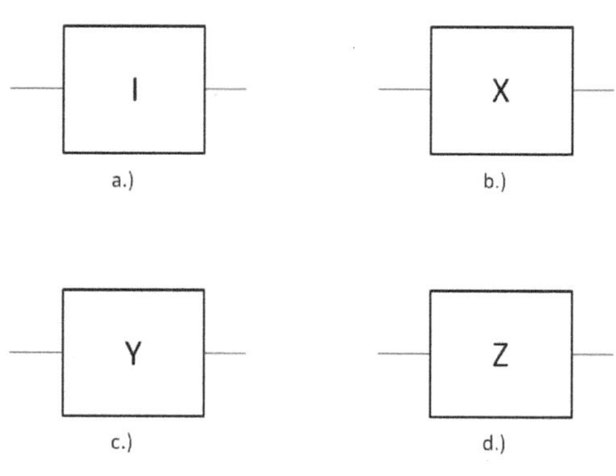

Fig. 5.2. Circuit depiction of the Pauli matrices as 1-qubit gates.

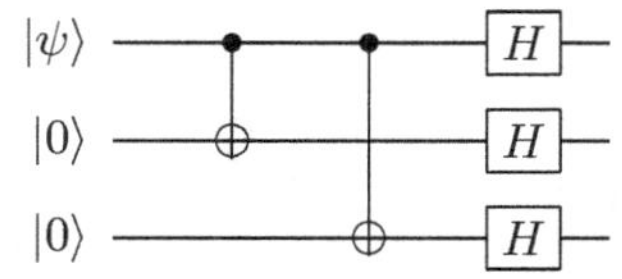

Fig. 5.3. Three-qubit QECC which corrects phase flip errors on a single qubit.

matrices, which means that for any Pauli matrix P, $UPU^\dagger = P'$, where U is a member of the Clifford group and P' is any Pauli matrix, including P. In fact, the $CNOT$, phase, and Hadamard matrices are the generators of the group, which means that some combination of these matrices can generate any other member of the Clifford group (up to global phase factors).

It has been shown that n-qubit states stabilized by members of the Clifford group are representable by only n stabilizers, an asymptotic improvement over the number of stabilizers required to represent arbitrary states [36]. The redundancy exploited here is the fact that applying a restricted set of operators to the stabilizer representation simply transforms the component matrices of the stabilizers to other matrices in the same finite group as the original stabilizers and operators. As will be shown in the next two sections, this fact allows us to avoid explicit calculation of matrix multiplications when applying operators and performing qubit measurement.

Before exploring these improvements, it is worth noting the relevance of the restricted class of circuits represented by Clifford group stabilizers. The X operator, for example, implements a bit flip in the computational basis, and similarly the Z operator implements a phase flip [61]. The matrices representing both of these operators are in the Clifford group and thus can be modeled directly using the stabilizer formalism. Importantly, it has been shown that arbitrary 1-qubit errors may be represented by linear combinations of Pauli matrices [36]. A useful related set of circuits that can be represented by Clifford group matrices are quantum error-correcting codes (QECC) [87], which will be revisited in Section 8.4. An example of a simple 3-qubit QECC is shown in Figure 5.3. This particular code uses extra qubits to detect a phase flip on a single qubit [88].

5.3 Data Structures

Symbolic names or labels for the Clifford group operators that make up stabilizers contain essentially the same information as the corresponding matrices. The most compact data structure for representing the stabilizer labels is a bit table [36, 1]. Consider the case of a stabilizer composed from the four Pauli operators $\{I, X, Y, Z\}$ defined by Equation 5.1. Each component requires two bits to represent it, so two binary variables, x_{ij} and z_{ij} are assigned to the Pauli operators. By convention, $x_{ij} = 1$ indicates the presence of an X, whereas $z_{ij} = 1$ indicates the presence of a Z. A value of 1 for both bits indicates the presence of a Y, whereas a value of 0 for both bits indicates the presence of an I [61]. For an n-qubit circuit, the table of bits takes the form

$$\left[\begin{array}{ccc|ccc} x_{00} & \cdots & x_{0(n-1)} & z_{00} & \cdots & z_{0(n-1)} \\ \vdots & \ddots & \vdots & \vdots & \ddots & \vdots \\ x_{(n-1)0} & \cdots & x_{(n-1)(n-1)} & z_{(n-1)0} & \cdots & z_{(n-1)(n-1)} \end{array}\right]$$

Each row denotes one stabilizer, so the dimensions of the table are $n \times 2n$. Continuing with the $|00\rangle$ state example, the corresponding stabilizers ZI and IZ have the following bit-table representation:

$$\left[\begin{array}{cc|cc} 0 & 0 & 1 & 0 \\ 0 & 0 & 0 & 1 \end{array}\right]$$

As noted earlier, the stabilizer formalism can incorporate gates outside of the Clifford group. It is easily shown that any 1-qubit operator can be decomposed into a sum of Pauli operators as follows:

$$U = \Sigma_j c_j P_j$$

where c_j is a complex-valued coefficient and P_j is a Pauli matrix. As a result, applying subsequent gates to U can be achieved by applying the gates to each Pauli matrix component. For example, consider the $\pi/8$ gate whose matrix T is

$$T = \begin{bmatrix} 1 & 0 \\ 0 & \frac{1+i}{\sqrt{2}} \end{bmatrix} = \alpha Z + \beta I$$

where $\alpha = 0.1464 - 0.3568i$ and $\beta = 1 - \alpha$. Suppose T is applied to a single qubit whose stabilizer is X. Hence, $TXT^\dagger$ produces $(X + Y)/\sqrt{2}$. The new state can be represented with two stabilizer symbols X and Y, and one or two coefficients. In the worst case, each application of a non-Clifford group operator quadruples the number of stabilizer symbols and coefficients that must be maintained per stabilizer, leading to asymptotically exponential runtime and memory usage. However, future implementations of quantum computers are expected to make heavy use of error-correcting codes, which means that many stabilizer gates will be applied for every non-stabilizer gate [61]. These techniques therefore will likely be heavily used for simulating a wide variety of sub-circuits.

5.4 Algorithms

With the bit-table data structure of the stabilizer formalism in place, the algorithms to manipulate the bit tables may now be explored. Explicit multiplication is not required to update the state after applying a gate that is a Clifford generator. Instead, a simple look-up table can be employed, which contains the rules for symbolically applying Clifford group generators to Pauli matrices [36, 1, 61]. These rules are given in Table 5.1. Each corresponds to the expression $Output = Gate * Input * Gate^{\dagger}$. Some transformations are not shown explicitly since they can be generated by combinations of the listed transformations. For instance, Y is equivalent to $SXS^{\dagger}$.

Gate	Input	Output
H	X	Z
	Z	X
S	X	Y
	Z	Z
$CNOT$	$X \otimes I$	$X \otimes X$
	$I \otimes$	$I \otimes X$
	$Z \otimes I$	$Z \otimes I$
	$I \otimes Z$	$Z \otimes Z$

Table 5.1. Transformation rules for applying Clifford group generators to Pauli operators.

Interestingly, the states produced by the stabilizer formalism define a limited set of probabilistic outcomes for the qubits. In particular, the probability of obtaining a $|0\rangle$ or $|1\rangle$ upon measurement of any qubit in such a state is always either 0, 1 or 1/2 [1]. Determining the measurement probability for a qubit, along with the transformation on the stabilizers induced by the measurement outcome, can be easily implemented by rules similar to those in Table 5.1 [1].

To illustrate these algorithms, we revisit the $|00\rangle$ example from the previous section and apply the circuit shown in Figure 4.5 to transform the state into an EPR pair, as described in Section 4.2. According to Table 5.1, applying a Hadamard gate to the first qubit transforms a Z to an X. In the table, this is accomplished by simply swapping the bits x_{11} and z_{11}. In general, a Hadamard gate is applied to qubit j by swapping the bits x_{ij} and z_{ij} for all i. All other transformations are accomplished by similar bit manipulations. As a result, any gate that is a Clifford group generator can be applied with runtime complexity $O(n)$. Using a modest number of extra bits, determining the measurement outcome and modifying the stabilizers based on that outcome can be done with runtime complexity $O(n^2)$ [1]. Thus, simulation of quantum circuits containing gates that are Clifford group generators requires $O(n^2)$ runtime and memory resources. To apply a gate that is outside the Clifford group, the bit table is copied and assigned a coefficient, as described earlier. Each table is then modified separately using the same bit manipulation rules. In the worst case, an exponential number of tables will be created.

The pseudocode for the bit-table implementations of CNOT, Hadamard, phase, and measurement updates are shown in Figures 5.4 and 5.5. The bit table used is similar to that in Equation 5.2, but augmented with $n+1$ extra rows so that the runtime complexity of measurement is $O(n^2)$ for n qubits. Without the extra rows, if the measurement outcome is deterministic, the runtime for measurement is on the order of n^3 in practice [1].

$$\left[\begin{array}{ccc|ccc|c} x_{00} & \cdots & x_{0(n-1)} & z_{00} & \cdots & z_{0(n-1)} & r_0 \\ \vdots & \ddots & \vdots & \vdots & \ddots & \vdots & \vdots \\ x_{(n-1)0} & \cdots & x_{(n-1)(n-1)} & z_{(n-1)0} & \cdots & z_{(n-1)(n-1)} & r_{(n-1)} \\ \hline x_{n0} & \cdots & x_{n(n-1)} & z_{n0} & \cdots & z_{n(n-1)} & r_n \\ \vdots & \ddots & \vdots & \vdots & \ddots & \vdots & \vdots \\ x_{(2n)0} & \cdots & x_{(2n)(n-1)} & z_{(2n)0} & \cdots & z_{(2n)(n-1)} & r_{(2n)} \end{array}\right]. \tag{5.2}$$

Revisiting the $\pi/8$ gate example, it is worth noting that each doubling of stabilizer symbols and coefficients requires duplication of the bit table being operated on for each Pauli matrix component of U. Subsequent application of Clifford group generators and measurement to each new bit table uses the same algorithms just described, albeit they are applied to each bit table; this links the exponential increase in memory usage to an exponential increase in runtime. The only minor change is that when performing measurement, each set of stabilizer symbols in a sum contributes a probability of measurement of 0, 1 or 1/2 multiplied by the appropriate coefficient $|c_j|^2$.

The foregoing discussion implies that simulating an n-qubit stabilizer circuit using a tabular representation of stabilizer states can be accomplished with memory and runtime whose complexity is $O(n^2)$ [1]. Relevant algorithms have been implemented by Aaronson in the CHP simulator, which is available in source code at [2]. Recently, Anders and Briegel [5] have developed a very memory-efficient simulation algorithm for stabilizer circuits that employs graphs instead of tables to represent stabilizer states. This work identifies a particular type of redundancy present in stabilizer tables, which can be exploited by specialized data structures and algorithms. The GraphSim simulator developed by Anders and Briegel only needs memory and runtime of $O(n \log n)$ in typical cases and was used to simulate circuits involving many thousands of qubits. This impressive development illustrates another aspect of quantum circuit simulation — by focusing on *typical inputs*, one may develop algorithms that are more efficient in practice, but not necessarily more efficient in the worst case.

5.5 Summary

A quantum circuit simulator can be viewed as implementing an abstract interface that describes typical operations performed on quantum gates and circuits. The particular implementation, be it physical with actual qubits or computational with ordinary bits on a classical computer, is required to produce the same results on the given input as a quantum circuit should. The output of a simulation may be the result of measuring a state, particularly in the case of a physical implementation, or a mathematical representation of a final state or operator prior to measurement, as in the case of classical simulation. For the sake of efficiency, classical implementations tend to restrict their inputs to sets of states, gates and circuits with some type of exploitable structure. A variety of data structures are employed to represent such properties in a compact way, and the corresponding algorithms manipulate the data structures directly.

The special technique for simulating stabilizer circuits illustrates some components of a general quantum simulator. The software-development aspect quantum simulation is illustrated by simulators written by Aaronson [2] and, more recently, Anders and Briegel [5]. From the application standpoint, the technique in [2] has been used to calculate fault-tolerance thresholds for individual gates in an ion trap architecture [57]. In principle, physicists building a system that implements such an architecture could use stabilizer circuit simulation to benchmark the reliability of individual components of the target system. The next chapter addresses non-stabilizer circuits and introduces more simulation algorithms.

```
CNOT_stabilizer(Xs, Zs, Rs, n) {
  for (a = 0; a < 2 * n; a + +) {
    for (b = 0; b < 2 * n; b + +) {
      for (i = 0; i < 2 * i; i + +) {
        Rs[i] = Rs[i] ⊕ Xs[i][a]&Zs[i][b]&(Xs[i][b] ⊕  Zs[i][a] ⊕ 1)
        Xs[i][b] = Xs[i][b] ⊕ Xs[i][a]
        Zs[i][a] = Zs[i][a] ⊕ Zs[i][b]
      }
    }
  }
}

Hadamard_stabilizer(Xs, Zs, Rs, n) {
  for (a = 0; a < 2 * n; a + +) {
    Rs[i] = Rs[i] ⊕ Xs[i][a]&Zs[i][a]
    tmp = Xs[i][a]
    Xs[i][a] = Zs[i][a]
    Zs[i][a] = tmp
  }
}

Phase_stabilizer(Xs, Zs, Rs, n) {
  for (a = 0; a < 2 * n; a + +) {
    Rs[i] = Rs[i] ⊕ Xs[i][a]&Zs[i][a]
    Zs[i][a] = Zs[i][a] ⊕ Xs[i][a]
  }
}

Rowsum(Xs, Zs, Rs, h, i) {
  sum = 2 * Rs[h] + 2 * Rs[i]
  for (j = 0; i < n; j + +) {
    g = 0
    g = ((Xs[i][j] == 0) && (Zs[i][j] == 0)) ? 0 : g
    g = (Xs[i][j] && Zs[i][j]) ? Zs[h][j] − Xs[h][j] : g
    g = (Xs[i][j] && (Zs[i][j] == 0)) ? Zs[h][h] * (2 * Xs[h][j ] − 1) : g
    g = ((Xs[i][j] == 0) && Zs[i][j]) ? Xs[h][j] * (1 − 2 * Zs[ h][j]) : g
  }
  sum + =  g
  Rs[h] = (sum%4 == 0) ? 0 : 1
  for (j = 0; j < n; j + +) {
    Xs[h][j] = Xs[i][j] ⊕ Xs[h][j]
    Zs[h][j] = Zs[i][j] ⊕ Zs[h][j]
  }
}
```

Fig. 5.4. Pseudocode for applying the CNOT, Hadamard, and phase operators using the stabilizer bit-table technique on qubits a and b (adapted from [1]).

```
Measure_stabilizer(Xs, Zs, Rs, n, a) {
  p = −1
  for (i = n; i < 2 ∗ n; i + +) {
    if (Xs[i][a]) {
      p = i
      break
    }
  }
  if (p > 0) {
    // Random measurement case
    for (i = 0; i < 2 ∗ n; i + +) {
      if ((i ! = p) && Xs[i][a]) {
        Rowsum(Xs, Zs, Rs, i, p)
      }
    }
    for (i = 0; i < n; i + +) {
      Xs[p − n][i] = Xs[p][i]
      Zs[p − n][i] = Zs[p][i]
      Xs[p][i] = Zs[p][i] = 0
    }
    Zs[p][a] = 1
    Rs[p] = 0
    if (Rand(1) > 0.5) {
      Rs[p] = 1
    }
    return Rs[p]
  }
  // Deterministic measurement case
  for (i = 0; i < n; i + +) {
    Xs[2n][i] = Zs[2n][i] = 0
  }
  Rs[2n] = 0
  for (i = 0; i < n; i + +) {
    if (Xs[i][a]) {
      Rowsum(Xs, Zs, Rs, 2n, i + n)
    }
  }
  return Rs[2n]
}
```

Fig. 5.5. Pseudo-code for measurement using the stabilizer bit-table technique on qubits a and b (adapted from [1]). Rand(x) generates a pseudo-random floating-point value between 0 and x.

6

Generic Circuit Simulation Techniques

In this chapter, we discuss several distinct methods that have been proposed for simulating arbitrary quantum circuits. They include qubit-wise multiplication, p-blocked simulation, the use of tensor networks, Vidal's "slightly entangled" technique, and a few others. Each approach has advantages that makes it especially well-suited to certain specific classes of quantum circuits. We also consider programming environments for quantum computing, most of which include front-ends to general-purpose quantum simulation software. Here we focus on the more powerful and sophisticated simulation algorithms and assess their advantages and disadvantages. This chapter may be skipped without significant loss of continuity.

6.1 Qubit-wise Multiplication

Recall that quantum gates are typically represented by matrices and quantum states by vectors. The action of a gate on a state then corresponds to matrix-vector multiplication. One popular circuit simulation technique is to implement the action of a k-input quantum gate on an n-qubit state vector ($k \leq n$) without explicitly storing a $2^n \times 2^n$-matrix representation [14, 62]. The basic idea is to simulate the full-fledged matrix-vector multiplication by a series of simpler operations. To illustrate, consider simulating a quantum circuit in which a 1-qubit Hadamard gate is applied to the third qubit of $|00100\rangle$, which comes from a state space containing 2^5 elements. A naive implementation is to construct a $2^5 \times 2^5$ matrix of the form $I \otimes I \otimes H \otimes I \otimes I$ and then multiply this matrix by the state vector. However, rather than compute $(I \otimes I \otimes H \otimes I \otimes I)|00100\rangle$, one can simply compute $|00\rangle \otimes H|1\rangle \otimes |00\rangle$, which produces the same result using a 2×2 matrix H. The same technique can be applied when the state space is in a superposition, such as $\alpha|00100\rangle + \beta|00000\rangle$. In this case, to simulate the application of a 1-qubit Hadamard gate to the third qubit, one can compute $|00\rangle \otimes H(\alpha|1\rangle + \beta|0\rangle) \otimes |00\rangle$. As in the previous case, a 2×2 matrix is sufficient.

While the above method allows one to compute a state space symbolically, in a realistic simulation environment state vectors may be much more complicated. Shortcuts that take advantage of the linearity of matrix-vector multiplication are desirable. For example, a single qubit can be manipulated in a state vector by extracting a certain set of two-dimensional vectors. Each vector in such a set is composed of two probability amplitudes. The corresponding qubit states for these amplitudes differ in value at the position of the qubit being operated on, but agree in every other qubit position. The two-dimensional vectors are then multiplied by matrices representing single qubit gates in the circuit being simulated. We refer to this technique as *qubit-wise multiplication* because the state space is manipulated one qubit at a time.

Obenland implemented a technique of this kind as part of an early simulator for quantum circuits [62]. His method applies one- and two-qubit operator matrices to state vectors of size 2^n. Unfortunately, in the best case where $k = 1$, this only reduces the runtime and memory complexity from $O(2^{2n})$ to $O(2^n)$, which is still exponential in the number of qubits. Another implicit limitation of Obenland's implementation is that it simulates with the state-vector representation only. The qubit-wise technique has been extended to enable density matrix simulation by Black et al. and is implemented in NIST's QCSim simulator [14]. As in its predecessor simulators, the arrays representing density matrices in QCSim tend to grow exponentially. This asymptotic bottleneck is demonstrated experimentally later (Section 8.4).

To illustrate the underlying steps of qubit-wise multiplication, consider a 1-qubit example in which a Hadamard gate is used to put a qubit into an equal superposition, $H|0\rangle = (|0\rangle + |1\rangle)/\sqrt{2}$. It is instructive to first represent the operation using matrices annotated with binary indices.

$$\begin{array}{c} \quad 0 \qquad 1 \\ \begin{bmatrix} \frac{1}{\sqrt{2}} & \frac{1}{\sqrt{2}} \\ \frac{1}{\sqrt{2}} & -\frac{1}{\sqrt{2}} \end{bmatrix}_{\substack{0\\1}} \end{array} \begin{bmatrix} 1 \\ 0 \end{bmatrix}_{\substack{0\\1}} = \begin{bmatrix} \frac{1}{\sqrt{2}} \\ \frac{1}{\sqrt{2}} \end{bmatrix}_{\substack{0\\1}}$$

The index being summed over in the Hadamard matrix is the column index, while in the state vector it is the row index. With the indices represented in binary, the direct summation becomes apparent. For each row index in the Hadamard matrix, sum over the column indices using a dot product that maps each column entry to each row entry in the state vector. Assuming that H represents the matrix array for the Hadamard operator and S represents the vector array for the state vector, the calculation of the first element becomes $H[0][0]*S[0]+H[0][1]*S[1]$, while that of the second element becomes $H[1][0]*S[0]+H[1][1]*S[1]$. The notation $A[i][j]$ refers to the ith row and jth column of A. This equivalent way of viewing the operation leads to an alternative implementation that performs dots products over various combinations of elements using the index bits being summed over. The general algorithm is simply to loop over all index bits in both operands not being summed over,

```
Qubit_wise_multiply(Op, S, qubit, gate_size) {
  gap = 1 << qubit
  gate_factor = 1 << gate_size
  num_groups = Floor(Rows(S)/gate_factor)
  sub_index = 0
  p_factor = 0
  a = 0
  Temp_vec = Vector(gate_factor)
  res = Vector(Rows(S))
  for (group_index = 0; group_index < num_groups; group_index++) {
    sub_index = group_index%gap
    if (sub_index == 0) {
      p_factor = gate_factor * group_index
      a = p_factor
    } else {
      a = p_factor + (group_index%gap)
    }
    for (index = 0; index < group_factor; index++) {
      Temp_vec[index] = S[a + gap * index]
    }
    res[a : a + gap * (index - 1) : gap] = Op * Temp_vec
  }                                                              }
  return res
}
```

Fig. 6.1. The k-qubit operator version of qubit-wise multiplication.

and for each set of bits perform a dot product over all index bits being summed over.

A pseudocode description of the full qubit-wise multiplication algorithm appears in Figure 6.1. Here `qubit` is the index number for the qubit affected by operator `Op`, `S` is the state vector array, `gate_size` is the size of `Op` in qubits, $<<$ is the left bit-shift operator, and `Vector` creates an empty vector array of the given size. The notation $X[a:b:c]$ refers to a slice of vector X starting at location a, ending at location b (inclusive), and an increment value or skip value of c. *Floor* returns the truncated integer form of a floating-point value.

6.2 P-blocked Simulation

The exponential complexity of state vectors or density matrices can be avoided by separating states into pieces that do not grow exponentially in size, wherever this is possible. To track such *separable states*, it is typical to employ entanglement metrics and devise state representations whose size is sensitive to these metrics. To this end, Jozsa and Linden studied states that are termed *p-blocked* if no subset of $p+1$ qubits are entangled. More generally, the set of all qubits is partitioned into k blocks $B_1, B_2, \ldots, B_k$. Then we can express the overall density matrix as:

$$\rho = \rho_1 \otimes \rho_2 \otimes \cdots \otimes \rho_k$$

where ρ_i is the density matrix for block B_i.

When k is large or equivalently, p is small, the complexity of simulation may be reduced by taking advantage of p-blocking [44]. Since each block requires at least 2^p coefficients, the space complexity grows with the number of entangled qubits. When $k = 1$, that is, no decomposition is possible, then no advantage is obtained from the p-blocking viewpoint. Operators that affect qubits within a single block can be applied directly to the block's density matrix using qubit-wise multiplication.

In general, however, an arbitrary circuit will likely contain gates operating on more than two qubits for which one of the affected qubits will be in a block B_i while other affected qubits will be in the next block B_{i+1}, and potentially in even more blocks if larger gates are used. Simulating gate operators that straddle several blocks requires combining all affected blocks via the tensor product into one large block. Once a single block is created from the smaller blocks, qubit-wise multiplication can be applied as before to the affected qubits of the density matrix for the entire block.

Keeping the large block intact is not ideal since simulation complexity usually grows exponentially in the number of block-straddling gates. The trick to avoiding this problem is to compute all partial traces between each contiguous partition of qubits in the large block. In general, the partial trace is not the inverse of the tensor product, particularly when entanglement has been introduced. However, if the tensor product of density matrices produced by partial tracing reproduces the original density matrix that the partial traces were taken from, then it *is* the inverse operation. Thus, by taking each of the combinatorially many partitions of qubits created by partial tracing, tensoring their density matrices together, and checking if the result is equal to the original large block, luck may have it that the simulation finds a new set of blocks that break up the large block and reduce the simulation complexity. If none of the partitions can reproduce the large block, then the large block must remain intact, thereby increasing the simulation complexity.

Although it may be possible to perform fewer partial traces with careful analysis of the given circuit, a more significant drawback is that for commonly used states such as $|\psi_{EPR}\rangle = (|00\ldots0\rangle + |11\ldots1\rangle)/\sqrt{2}$, known as the EPR, Bell or "cat" state, the p-block representation requires space that is exponential in n when $p \gg \log(n)$. For example, consider a 2-qubit version of the foregoing EPR state. Computing the partial trace over both qubits produces two density matrices whose tensor product is not equal to the density matrix of the original state, as shown below.

$$tr_{|\psi_A\rangle\langle\psi_A|}(|\psi_{EPR}\rangle\langle\psi_{EPR}|) = tr\left(\begin{bmatrix} \frac{1}{2} & 0 & 0 & \frac{1}{2} \\ 0 & 0 & 0 & 0 \\ 0 & 0 & 0 & 0 \\ \frac{1}{2} & 0 & 0 & \frac{1}{2} \end{bmatrix}\right) = tr\begin{bmatrix} \frac{1}{2} & 0 \\ 0 & \frac{1}{2} \end{bmatrix}$$

$$\begin{bmatrix} \frac{1}{2} & 0 \\ 0 & \frac{1}{2} \end{bmatrix} \otimes \begin{bmatrix} \frac{1}{2} & 0 \\ 0 & \frac{1}{2} \end{bmatrix} = \begin{bmatrix} \frac{1}{4} & 0 & 0 & 0 \\ 0 & \frac{1}{4} & 0 & 0 \\ 0 & 0 & \frac{1}{4} & 0 \\ 0 & 0 & 0 & \frac{1}{4} \end{bmatrix} \neq |\psi_{EPR}\rangle\langle\psi_{EPR}|$$

As a result, p must be increased upon creation of the EPR pair, and p increases exponentially with the number of qubits entangled in this fashion when creating an n-qubit EPR state. In contrast, QuIDDs can represent $|\psi_{EPR}\rangle$ using $O(n)$ space by exploiting the massive redundancy in the amplitudes of this state, as will be demonstrated in Chapters 7 and 8.

6.3 Tensor Networks

While p-blocked simulation targets separable states during quantum simulation, tensor networks capture structure in the circuits that allows efficient simulation. This idea was first studied in the physics community using such new concepts as Matrix Product States (MPS) and Projected Entangled Pairs States (PEPS) [105, 113, 65] In the computer science community, Markov and Shi developed a self-contained circuit simulation algorithm [55] in terms of networks of tensors, which are multi-dimensional generalizations of matrices. In this approach, tensors represent density matrix states and operators. For example, the operator U acting on a input qubits and b output qubits is denoted as follows [55].

$$[U_{\sigma_1,\sigma_2,\ldots,\sigma_a,\tau_1,\tau_2,\ldots,\tau_b}]_{\sigma_1,\sigma_2,\ldots,\sigma_a,\tau_1,\tau_2,\ldots,\tau_b}$$

where each $\sigma_i, \tau_j \in |b_1\rangle\langle b_2| : b_1, b_2 \in \{0,1\}$. Here each index can take on one of the four possible index values of a 1-qubit density matrix.

A separate tensor is created for each gate in a circuit. Each tensor is treated like a node in a graph, and is connected to other tensors via shared qubit indices, which correspond to input/output connections. These graphs or *tensor networks* make use of an operation called *tensor contraction* which merges connected nodes containing tensors into a single tensor. Tensor contraction is a key operation used not only in tensor networks but also in two other simulation techniques discussed later.

The contraction itself is simply the multi-dimensional generalization of the dot product and matrix multiplication. As an example, consider the tensor contraction of tensor g and h over a shared output/input connection [55]:

$$f_{i_1,\ldots,i_m,j'_1,\ldots,j'_{n'}} = \sum_{j_1,\ldots,j_n} g_{i_1,\ldots,i_m,j_1,\ldots,j_n} \cdot h_{j_1,\ldots,j_n,j'_1,\ldots,j'_{n'}}$$

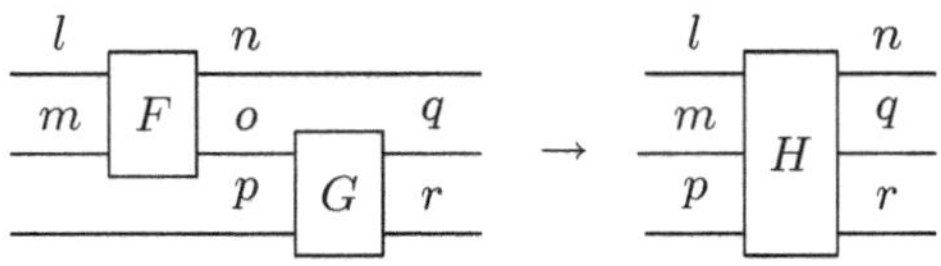

Fig. 6.2. Tensor contraction of shared wire (index) o for tensors F and G, each of which represents a 2-qubit gate.

The goal of this method is to contract all tensors into a single tensor describing the action of the circuit on qubits of interest. Depending on how the tensors are connected, contractions may either decrease, increase, or leave unaffected the dimensions of the resultant tensor as compared to the dimensions of the separate tensors. To illustrate, consider the tensor contraction of F and G over the shared index o, as shown in Figure 6.2. F and G are tensor representations of two 2-qubit quantum gates where an output wire (output tensor index) o of F is an input wire (input tensor index) to G. Notice that in this case, tensor contraction produces a new tensor H with larger dimensions than F and G.

Figure 6.3 provides pseudocode for the tensor contraction algorithm. Simulation with this method is exponential in d, the maximum dimension of any tensor created by contractions. The tensor-network approach is also applicable to instances of one-way quantum computation [19]. However, tensor networks can be insensitive to separable states, and therefore, in practice, are combined with other simulation techniques [81]. Aharonov et al.[4] as well as Yoran and Short [112] employ the tensor-network approach to approximately simulate the quantum Fourier transform (QFT) — a key operation in Shor's quantum factoring algorithm [82]. They first simplify well-known QFT circuits by omitting a small subset of gates so that the resulting transform remains very close to QFT, but the circuit's treewidth becomes so small that tensor contractions complete simulation in polynomial time. However, this result does not enable efficient number factoring on a classical computer because a different operation called modular exponentiation remains a bottleneck when applied to a superposition of inputs.

By making use of tensor contraction, the qubit-wise multiplication algorithm in Section 6.1 can be described very compactly, as shown in Figure 6.4. An n-qubit state vector is indexed by n indices, each having a range of two thus: $|\psi\rangle = S^{i_0 i_1 \cdots i_{n-1}}$. A 1-qubit operator applied to qubit a takes the form $U^p_{i_a}$, while a 2-qubit operator applied to qubits a and b takes the form $V^{pq}_{i_a i_b}$. Generalizing to the k-qubit operator case merely requires adding more indices to the operator representation. The density-matrix version is similar except that S will incorporate superscripts and subscripts, and an extra tensor contraction will be performed to multiply by the complex-conjugate transpose of the operator.

```
Contract(A, B, shared_indices_A, shared_indices_B) {
  num_shared_bits = Size(shared_indices_A)
  num_row_bits_A = Round(Log(Rows(A)))
  num_col_bits_A = Round(Log(Cols(A)))
  num_row_bits_B = Round(Log(Rows(B)))
  num_col_bits_B = Round(Log(Cols(B)))
  shared_mask_A = 0
  for (curr_bit = 0; curr_bit < num_shared_bits; curr_bit++) {
    shared_mask_A |= 1 << (num_row_bits_A - shared_indices_A[curr_bit] - 1)
  }
  shared_mask_B = 0
  for (curr_bit = 0; curr_bit < num_shared_bits; curr_bit++) {
    shared_mask_B |= 1 << (num_col_bits_B - shared_indices_B[curr_bit] - 1)
  }
  sum_terms = 1 << Size(shared_indices_A)
  res_row_bits = num_row_bits_B + num_row_bits_A - num_shared_bits
  res_col_bits = num_col_bits_A + num_col_bits_B - num_shared_bits
  row_dim = 1 << res_row_bits
  col_dim = 1 << res_col_bits
  for (i = 0; i < row_dim; i++) {
    for (j = 0; j < col_dim; j++) {
      res[i][j] = 0
      for (k = 0; k < sum_terms; k++) {
        row_A = curr_bit_i = curr_bit_k = 0
        for (curr_bit = 0; curr_bit < num_row_bits_A; curr_bit++) {
          if (shared_mask_A & (1 << curr_bit)) {
            row_A |= (k & (1 << curr_bit_k)) ? 1 << curr_bit : 0
            curr_bit_k++
          }
          else {
            row_A |= (i & (1 << (curr_bit_i + num_row_bits_B))) ? 1 << curr_bit : 0
            curr_bit_i++
          }
        }
        col_B = curr_bit_j = curr_bit_k = 0
        for (curr_bit = 0; curr_bit < num_col_bits_B; curr_bit++) {
          if (shared_mask_B & (1 << curr_bit)) {
            col_B |= (k & (1 << curr_bit_k)) ? 1 << curr_bit : 0
            curr_bit_k++
          }
          else {
            col_B |= (j & (1 << curr_bit_j)) ? 1 << curr_bit : 0
            curr_bit_j++
          }
        }
        col_a = curr_bit_j = 0
        for (curr_bit = 0; curr_bit < num_col_bits_a; curr_bit++) {
          offset_j = num_col_bits_b - num_shared_bits
          col_A |= (j & (1 << (curr_bit + offset_j))) ? 1 << curr_bit : 0
        }
        row_B = curr_bit_i = 0
        for (curr_bit = 0; curr_bit < num_row_bits_B; curr_bit++) {
          row_B |= i & (1 << curr_bit)
        }
        res[i][j] += A[row_A][col_A] * B[row_B][col_B]
      }
    }
  }
  return res
}
```

Fig. 6.3. Tensor contraction over given shared indices. C-style bit-wise operations are used including & (AND), | (OR) and << (left bit-shift).

```
Qubit_wise_multiply(Op, S, Qubits) {
  return Contract(S, Op, Qubits)
}
```

Fig. 6.4. The k-qubit operator version of qubit-wise multiplication using tensor contraction. The array **Qubits** contains qubit indices affected by operator **Op**.

To illustrate this implementation in action, consider again the example in Section 6.1 of putting a single qubit into an equal superposition by means of a Hadamard gate H. Using tensor notation, the tensor contraction implementation has the form $H^p_{i_0} S^{i_0} = T^p$, where S and T are state vectors and p and i_0 are indices that take on values 0 or 1.

6.4 Slightly-entangled Simulation

The p-blocked simulation approach separates states using only the tensor product operation. More sophisticated decomposition techniques have been developed that exploit state separability even further. Vidal has devised one such technique, which utilizes the Schmidt decomposition (SD) of the quantum state [106]. EPR states and separable states are represented with only quadratic overhead by Vidal's technique [106]. Consider n qubits ordered from 0 to $n-1$. A *bipartite splitting* $A{:}B$ of the qubits is given by any integer k between 1 and $n-2$ in the sense that the two qubit sets with indices $i \leq k$ and $j > k$ form complementary partitions A and B. Then the state $|\psi\rangle$ can be decomposed as follows [30].

$$|\psi\rangle = \sum_{\alpha=0}^{\chi_A - 1} \lambda_\alpha \left|\Phi_\alpha^{[A]}\right\rangle \otimes \left|\Phi_\alpha^{[B]}\right\rangle \tag{6.1}$$

Here, $\left|\Phi_\alpha^{[A]}\right\rangle$ and $\left|\Phi_\alpha^{[B]}\right\rangle$ are two orthonormal bases, and $\sum_\alpha |\lambda_\alpha|^2 = 1$. It is common to take $\left|\Phi_\alpha^{[A]}\right\rangle$ and $\left|\Phi_\alpha^{[B]}\right\rangle$ as eigenvectors of the reduced density matrices $\rho^{[A]}$ and $\rho^{[B]}$, respectively, which both have the same eigenvalue $|\lambda_\alpha|^2 > 0$. The *Schmidt rank* χ_A is a measure of entanglement between partitions A and B, and each of χ_A addends consists of two vector terms. The entanglement of state $|\psi\rangle$ can be quantified by the maximum χ_A over all possible bipartite splittings $A : B$, $\chi \equiv \max_A \chi_A$ [106, Equation 2]. Depending on the amount of entanglement, χ can range from 1 for fully separable states, to 2^n for fully entangled states.

The space complexity of Vidal's representation and the time complexities of algorithms that use it are functions of χ. In particular, Vidal decomposes an n-qubit state into sums of tensor products [106, Equation 15] that we refer to as a *dense tensor decomposition* (DTED).

$$|\psi\rangle = \sum_{\alpha_0=0,\cdots,\alpha_{n-2}=0}^{(\chi_0-1),\cdots,(\chi_{n-2}-1)} \left|\varphi^{[0]}_{\alpha_0}\right\rangle \lambda^{[0]}_{\alpha_0} \left|\varphi^{[1]}_{\alpha_0\alpha_1}\right\rangle \ldots \lambda^{[n-2]}_{\alpha_{n-2}} \left|\varphi^{[n-1]}_{\alpha_{n-2}}\right\rangle$$

In this equation, vectors $\left|\varphi^{[l]}_{\alpha_{l-1}\alpha_l}\right\rangle$ are unnormalized 1-qubit states, and the Schmidt coefficients $\lambda^{[l]}_{\alpha_l}$ express the correlation information between qubits $0, 1, \ldots, \ell$ and qubits $l+1, l+2, \ldots, n-1$ (the tensor product symbols are omitted for simplicity). Each α_l index may range from 0 to $\chi_l - 1$. The DTED of a pure n-qubit state $|\psi\rangle$ is derived by applying the SD $n-1$ times to the bipartite splittings $0 : n-1, 1 : n-1, \ldots, n-2 : n-1$, in such a way that the maximal possible rank is χ. Each time, this process generates χ_l coefficients $\lambda^{[l]}_{\alpha_l}$ and χ_l^2 2-element vectors $\left|\varphi^{[l]}_{\alpha_{l-1}\alpha_l}\right\rangle$. Therefore, the DTED decomposes $|\psi\rangle$ into a sum of up to χ^n separable states and requires $n(2\chi^2+\chi)$ complex-valued coefficients [106].

To simulate 1- and 2-qubit gates, one uses algorithms that update the DTED state representation. Vidal gives algorithms that take $O(\chi^2)$ time for 1-qubit gates and $O(\chi^3 + n\chi^2)$ time [1] for nearest-neighbor 2-qubit gates [106]. For a generic circuit with g gates, Vidal's protocol runs in $O(ng\chi^3 + n^2 g\chi^2)$ time. In particular, applying 2-qubit operators to qubits l and $l+1$ requires solving a potentially large eigenvalue problem to update $\lambda^{[l]}_{\alpha_l}$ [106]. While the precise complexity of measurement is not given in [106], we believe that it requires $O(n\chi^2)$ time in the DTED formalism. Vidal notes that measurement can be accomplished in time polynomial in χ, but apparently assumes in the analysis that $\chi = \Omega(n)$.

To see how these complexity results are derived, consider the algorithms for 1- and 2-qubit operator updates in Vidal's DTED representation [106]. A 1-qubit unitary operator U is applied to qubit l as follows:

$$\left|\varphi'^{[l]}_{\alpha_{l-1}\alpha_l}\right\rangle = U \left|\varphi^{[l]}_{\alpha_{l-1}\alpha_l}\right\rangle \forall \alpha_l, \alpha_{l+1} = 0, \cdots, (\chi - 1)$$

This operation takes $O(\chi^2)$ time since $\alpha_{l-1}, \alpha_l \leq \chi$. DTED updates for 2-qubit operators applied to qubits l and $l+1$ are much more involved. Vidal explicitly solves for the eigenvalues and eigenvectors of $\rho^{[(l+1)..(n-1)]}$ (see Equation 6.1), which requires several major steps. The first step is to apply the 2-qubit unitary operator V to the substates corresponding to qubits l and $l+1$ in the following way.

$$\Theta^{[l,(l+1)]}_{\alpha_{l-1}\alpha_{l+1}} = \sum_{\alpha_l} V \left|\varphi^{[l]}_{\alpha_{l-1}\alpha_l}\right\rangle \lambda^{[l]}_{\alpha_l} \left|\varphi^{[l+1]}_{\alpha_l\alpha_{l+1}}\right\rangle \tag{6.2}$$

[1] When a 2-qubit operator is applied to qubits l and $l+1$, partial traces over all other qubits must be computed, requiring $O(n\chi^2)$ time for this step [106, Equations 13, 14, 18, 19, 23, and 26]. This term is not included in [106, Lemma 2] because it is dominated by χ^3 when $\chi \gg n$. However, it can be significant for slightly-entangled states which are the focus of [106].

The resultant density matrix of the second partition becomes

$$\rho'^{[(l+1)\cdot\cdot(n-1)]} = \sum_{j,j',\alpha_{l+1},\alpha'_{l+1}} \Big(\sum_{\alpha_{l-1}} \langle \alpha_{l-1} | \alpha_{l-1} \rangle \Theta^{[l,(l+1)]}_{\alpha_{l-1}\alpha_{l+1}} (\Theta^{[l,(l+1)]}_{\alpha_{l-1}\alpha'_{l+1}})^{*}\Big) |j\alpha_{l+1}\rangle \langle j'\alpha'_{l+1}| \tag{6.3}$$

where $j = 0, 1$. Also, following [106, Equations 18 and 19], we use

$$|\alpha_{l-1}\rangle \equiv \lambda^{[l-1]}_{\alpha_{l-1}} \left|\Phi^{[0\cdot\cdot(l-1)]}_{\alpha_{l-1}}\right\rangle \tag{6.4}$$

$$|\alpha_{l+1}\rangle \equiv \lambda^{[l+1]}_{\alpha_{l+1}} \left|\Phi^{[(l+2)\cdot\cdot(n-1)]}_{\alpha_{l+1}}\right\rangle$$

Computing $\langle \alpha_{l-1} | \alpha_{l-1} \rangle$ using Equation 6.4 requires $O(n\chi^2)$ time. Equation 6.2 is computed using $O(\chi^3)$ time since there are three consecutive α indices, each of which is bounded by χ. With $\langle \alpha_{l-1} | \alpha_{l-1} \rangle$ and Equation 6.2 computed, Equation 6.3 is computed using $O(\chi^3)$ time since it involves summing over all combinations of α_{l+1}, α'_{l+1}, and α_{l-1}. The new Schmidt coefficients $\lambda'^{[l]}_{\alpha_l}$ are generated by solving for the eigenvalues of $\rho'^{[(l+1)\cdot\cdot(n-1)]}$, which can be done using $O(\chi^3)$ time. The new states $\left|\varphi^{[l+1]}_{\alpha_l \alpha_{l+1}}\right\rangle$ are computed by decomposing the eigenvalues and eigenvectors in terms of $|j\alpha_{l+1}\rangle$ in $O(\chi)$ time. Lastly, the new states $\left|\varphi^{[l]}_{\alpha_{l-1}\alpha_l}\right\rangle$ are computed by decomposing the eigenvalues, eigenvectors, and $\langle \alpha_{l+1} | \alpha_{l+1} \rangle$ terms with respect to $|\alpha_{l-1} i\rangle$, where $i = 0, 1$, requiring $O(\chi)$ time. The overall time complexity of the 2-qubit operator update is therefore $O(\chi^3)$ while increasing α_l by up to 2χ. The pseudocode for updating 1- and 2-qubit operators is given in Figure 6.5.

To illustrate how Vidal's protocol works, consider again the creation of an EPR pair. A simplified version of the notation is used below to track Vidal's algorithms after application of the Hadamard and CNOT gates. Initially, the states are unentangled since both qubits have the value $|0\rangle$. This means that $\chi = 1$, $\left|\Phi^{[0]}_0\right\rangle = \left|\Phi^{[1]}_0\right\rangle = |0\rangle$, and $\lambda_0 = 1$.

$$H\,|0\rangle \otimes |0\rangle = |+\rangle \otimes |0\rangle$$

$$CNOT(|+\rangle \otimes |0\rangle) = \frac{1}{\sqrt{2}}(|0\rangle \otimes |0\rangle) + \frac{1}{\sqrt{2}}(|1\rangle + |1\rangle)$$

The Hadamard gate does not increase χ since it is applied only to the first qubit. The CNOT gate, however, increases χ by one, as indicated by the presence of a second tensor product term in the summation. The CNOT gate is applied by computing the tensor product $|+\rangle \otimes |0\rangle$, multiplying the resulting 4-element vector by the matrix representing CNOT, computing the density matrix of the resulting vector via the outer product, and solving for both the eigenvalues and eigenvectors of this density matrix. λ contains the square roots of the two eigenvalues shared by the reduced density matrices of each qubit

```
One_qubit_update(Op, Γ, l) {
  Shared_indices[0] = 0
  Contract(Γ[l], Op, Shared_indices, Shared_indices)
}

Two_qubit_update(Op, Γ, λ, l) {
  // Create Θ
  shared_indices_A[0] = 0
  shared_indices_B[0] = 2
  θ = Contract(Γ[l], Op, shared_indices_A, shared_indices_B)
  Theta = Contract(Γ[l + 1], θ, shared_indices_A, shared_indices_B)
  shared_indices_A[1] = 0
  shared_indices_B[1] = 1
  θ = Contract(λ[l], θ, shared_indices_A, shared_indices_B)
  // Create ⟨α|α⟩
  α = Gen_inner(Γ, λ, l − 1)
  // Create ρ'
  Clear(shared_indices_A)
  ρ = Contract(α, θ, shared_indices_A, shared_indices_A)
  ρ = Contract(ρ, Conj(θ), shared_indices_A, shared_indices_A)
  shared_indices_A[0..1] = {0, 0}
  shared_indices_B[0..1] = {4, 6}
  ρ = Contract(Ones(Size(α)), ρ, shared_indices_A, shared_indices_B)
  shared_indices_B[0..1] = {0, 2}
  ρ = Contract(Ones(2), ρ, shared_inidces_A, shared_indices_B)
  // Create λ^[l]' and Γ^[l+1]'
  Eigensolve(ρ, Eigenvalues, Eigenvectors)
  for (b = 0; b < Size(Eigenvalues); b++) {
    λ[l][b] = Sqrt(Eigenvalues[b])
    for (j = 0; j < 2; j++) {
      for (g = 0; g < Size(λ[l + 1]); g++) {
        Γ'[j][b][g] =Eigenvectors[b][j * Size(λ[l + 1]) + g]/λ[l + 1][g]
      }
    }
  }
  Γ[l + 1] = Γ'
  // Create ⟨γ|γ⟩
  γ = Gen_inner(Γ, λ, l + 1)
  // Create Γ^[l]'
  Clear(shared_indices_A)
  Γ' = Contract(Γ[l + 1], θ, shared_indices_A, shared_indices_A)
  Γ' = Contract(Γ', γ, shared_indices_A,shared_indices_A)
  shared_indices_B[0..2] = {0, 1, 2}
  shared_indices_A[0..2] = {0, 0, 0}
  Γ' = Contract(Ones(2), Γ', shared_indices_A, shared_indices_B)
  shared_indices_B[0..2] = {4, 6, 7}
  Γ' = Contract(Ones(Size(λ[l + 1])), Γ', shared_indices_A,shared_indices_B)
  Clear(shared_indices_A)
  Clear(shared_indices_B)
  shared_indices_A[0] = 0
  shared_indices_B[0] = 5
  Γ' = Contract(Ones(Size(λ[l − 1])), Γ', shared_indices_A,shared_indices_B)
  for (b = 0; b < Size(Eigenvalues); b++) {
    Γ'[i][a][b]/ = λ[l][b]
  }
  Γ[l] = Γ'
}
```

Fig. 6.5. 1- and 2-qubit operator updates for slightly-entangled simulation. `Ones(x)` creates a vector of dimension x populated with 1's `Eigensolve(A, V, D)` computes the eigenvalues and eigenvectors of matrix A and stores them in V and D, respectively.

(reduced via the partial trace), and the new state vectors are the eigenvectors of the reduced density matrices.

A potential drawback of DTED and Vidal's simulation protocol is the redundancy in state encoding. For a generic state with maximum entanglement ($\chi = 2^n$), DTED requires $\Omega(n2^{2n})$ coefficients, whereas 2^n amplitudes suffice to characterize the state. Interestingly, p-blocked simulation and QuIDDs represent generic, maximally entangled states using only $O(2^n)$ space [44, 99]. A key open question is whether the extra coefficients are necessary to ensure that the space and time complexity of quantum simulation remain polynomial in χ.

6.5 Summary

This chapter described a number of sophisticated techniques for quantum circuit simulation. Each one exploits a particular property of quantum states and/or operators to increase computation speed or reduce memory needs. Qubit-wise multiplication, for example, exponentially reduces the memory complexity of storing operators, but maintains state vectors and density matrices in their full form and as such does not reduce the runtime complexity of *applying* operators. Both p-blocked simulation and Vidal's technique decompose states into partitions that involve small amounts of entanglement. In the p-blocking method, states are separated into tensor products of density matrices of size at most $O(2^{2p})$. However, tensor products alone do not compress many forms of entanglement, making p an overestimate of the level of entanglement in the system. Vidal improves upon this measure of entanglement by basing state decomposition on the maximal Schmidt rank χ.

Besides the techniques covered, several other simulation techniques are known, but have so far have seen little use. Valiant described a technique which can efficiently simulate a class of quantum circuits whose properties are related to the Pfaffian of relevant matrices [91] — a mathematical construct resembling the determinant of a matrix. Unlike the stabilizer formalism discussed in the last chapter, the class of gates efficiently simulable by this technique covers few known practical applications. Several simulation techniques suffer similar limitations [37]. Other simulation techniques, including MATLAB's "packed" representation, apply data compression to matrices and vectors, but cannot perform matrix-vector multiplication without first decompressing the matrices and vectors.

In the remaining chapters, we will build a full simulator by taking a decision diagram technique (QuIDD simulation) and turning it into a practical tool for a variety of quantum circuit CAD problems. The theoretical foundations of QuIDD-based simulation will be explained along with various practical implementation and application details. Most of the issues addressed apply to quantum simulator development based on any generic approach.

7

State-Vector Simulation with Decision Diagrams

This chapter covers the theory behind the QuIDD and related data structures, all of which are based on the decision diagram (DD) concepts described in Section 2.2. As will be seen, QuIDDs allow simulations of n-qubit systems to achieve runtime and memory complexities that range from constant-time to exponential, and the worst case is not typical for structured circuits arising in practical applications. Later in the chapter, benchmark results with Grover's quantum search algorithm will reveal that a QuIDD-based simulator outperforms several implementations of qubit-wise multiplication in terms of asymptotic runtime and memory usage.

7.1 Quantum Information Decision Diagrams

The QuIDD concept was born out of the observation that vectors and matrices arising in quantum computation contain entries and sub-matrices that occur repeatedly. This type of repeated sub-structure can be captured and exploited by certain BDD variants, such MTBDDs and ADDs,, which were introduced in Section 2.2.

A QuIDD can be viewed as an ADD with the following properties:

1. The data values associated with terminal nodes are complex numbers.
2. Rather than contain data values explicitly, the terminal nodes contain integer indices (pointers), which map to a separate array of complex numbers and greatly simplify the implementation of QuIDD operations.
3. The variable ordering of QuIDDs interleaves row and column variables, which favors compression of repeated sub-structures.
4. Vectors and matrices in the quantum domain have sizes that are powers of 2, so QUIDDs do not require padding with 0's, unlike general ADDs.

We will demonstrate using our QuIDD-based simulator QuIDDPro that these properties greatly enhance simulation performance for quantum computation.

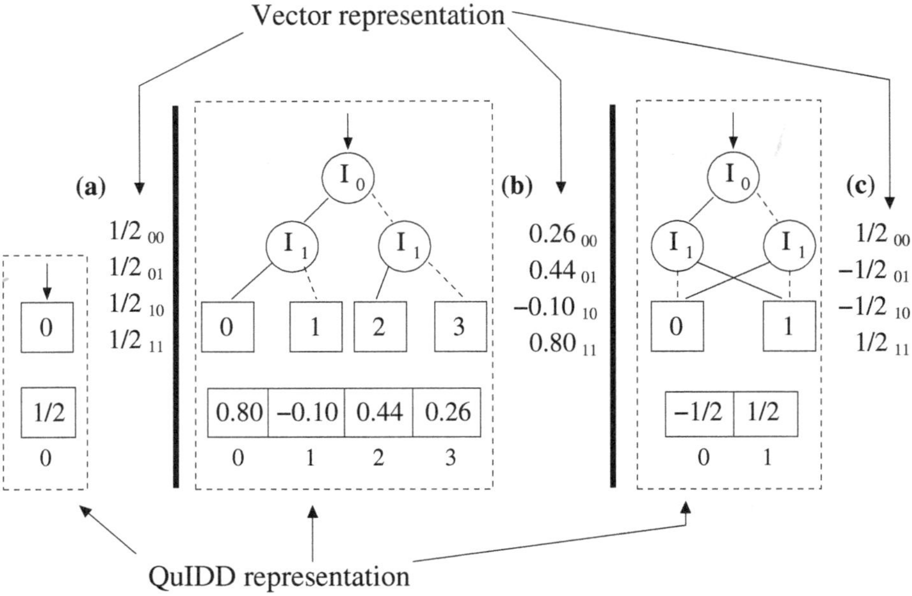

Fig. 7.1. Sample QuIDDs for state vectors of (a) best, (b) worst, (c) medium size.

Representing Vectors and Matrices

Figure 7.1 shows the QuIDD structure for three 2-qubit states. We consider the indices of the four vector elements to be binary numbers, and define their bits as decision variables of QuIDDs. A similar definition is used for ADDs [6]. For example, traversing the *then* edge (solid line) of node I_0 in Figure 7.1c is equivalent to assigning the value 1 to the first bit of the 2-bit vector index. Traversing the *else* edge (dotted line) of node I_1 in the same figure is equivalent to assigning the value 0 to the second bit of the index. These traversals bring us to the terminal value $-\frac{1}{2}$, which is precisely the value at index 10 in the vector representation.

QuIDD representations of matrices extend those of vectors by adding a second type of variable node and enjoy the same reduction rules and compression benefits. Consider the 2-qubit Hadamard matrix annotated with binary row and column indices shown in Figure 7.2a. In this case there are two sets of indices: the first (vertical) set corresponds to the rows, while the second (horizontal) set corresponds to the columns. We assign the variable name R_i and C_i to the row and column index variables, respectively. This distinction between the two sets of variables was originally noted in several works, including [6]. Figure 7.2b shows the QuIDD form of this sample matrix where it is used to modify the state vector $|00\rangle = (1, 0, 0, 0)$ via matrix-vector multiplication, an operation discussed in more detail later in this section.

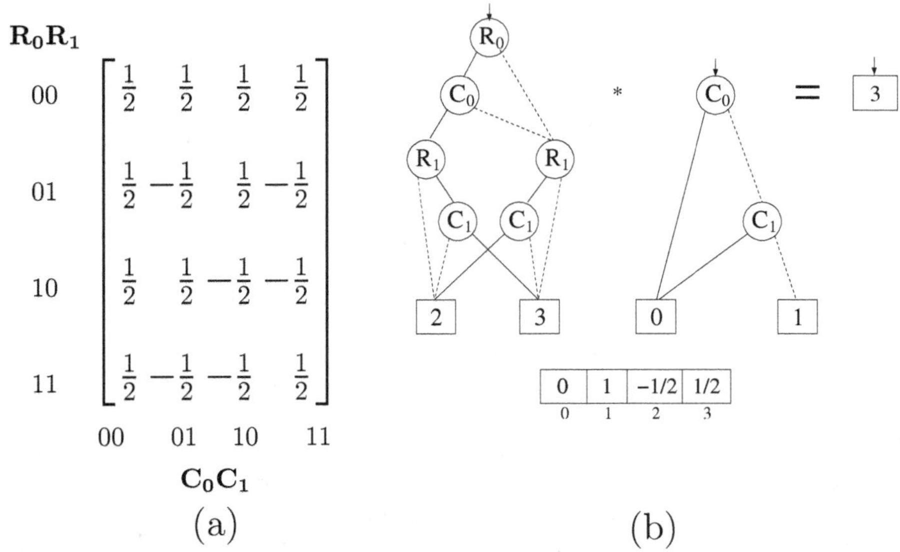

Fig. 7.2. (a) 2-qubit Hadamard matrix, and (b) its QuIDD representation multiplied by $|00\rangle = (1,0,0,0)$. Note that the vector and matrix QuIDDs share the entries in a terminal array that is global to the computation.

Variable Ordering. As explained in Section 2.2, variable ordering can drastically affect the level of compression achieved in BDD-based structures such as QuIDDs. The CUDD programming library [83], which is incorporated into QuIDDPro, offers sophisticated dynamic variable-reordering techniques that achieve performance improvements in various BDD applications. However, dynamic variable reordering has significant time overhead, whereas finding a good static ordering in advance is preferable in some cases. Good variable orderings are highly dependent upon the problem domain. In the case of quantum computing, we notice that all matrices and vectors contain 2^n elements where n is the number of qubits represented. Additionally, the matrices are square and non-singular [61].

McGeer et al. demonstrated that ADDs representing certain rectangular matrices can be operated on more efficiently if row and column variables are interleaved [26]. This interleaving employs the following variable ordering: $R_0 \prec C_0 \prec R_1 \prec C_1 \prec \ldots \prec R_n \prec C_n$. Intuitively, the interleaved ordering causes compression to favor regularity in particular sub-structures of the matrices that are partitions broken up into equally sized quadrants or blocks. We observe that such regularity is created by tensor products that allow multiple quantum gates to operate in parallel, or allow the number of inputs of a quantum gate to increase. The tensor product $A \otimes B$ multiplies each element of A by the whole matrix B to create a larger matrix, so by definition, the tensor product propagates block patterns in its operands.

To illustrate the notion of sub-structure and how QuIDDs take advantage of it, consider the tensor product of two 1-qubit Hadamard operators,

$$\left[\begin{array}{c|c} (\frac{1}{\sqrt{2}}) & (\frac{1}{\sqrt{2}}) \\ \hline (\frac{1}{\sqrt{2}}) & (-\frac{1}{\sqrt{2}}) \end{array}\right] \otimes \left[\begin{array}{c|c} (\frac{1}{\sqrt{2}}) & (\frac{1}{\sqrt{2}}) \\ \hline (\frac{1}{\sqrt{2}}) & (-\frac{1}{\sqrt{2}}) \end{array}\right] = \left[\begin{array}{c|c} \begin{pmatrix} \frac{1}{2} & \frac{1}{2} \\ \frac{1}{2} & -\frac{1}{2} \end{pmatrix} & \begin{pmatrix} \frac{1}{2} & \frac{1}{2} \\ \frac{1}{2} & -\frac{1}{2} \end{pmatrix} \\ \hline \begin{pmatrix} \frac{1}{2} & \frac{1}{2} \\ \frac{1}{2} & -\frac{1}{2} \end{pmatrix} & \begin{matrix} -\frac{1}{2} & -\frac{1}{2} \\ -\frac{1}{2} & \frac{1}{2} \end{matrix} \end{array}\right]$$

The above matrices have been separated into quadrants, each of which represents a block. For the Hadamard matrices depicted, three of the four blocks are equal in both of the 1-qubit matrices and also in the larger 2-qubit matrix (the equivalent blocks are surrounded by parentheses). This repetition of equivalent blocks demonstrates that the tensor product of two equal matrices propagates block patterns. In this example, all blocks but the lower-right block of an n-qubit Hadamard operator are equal. Furthermore, the structure of the 2-qubit matrix implies a recursive sub-structure, which can be seen by recursively partitioning each of the quadrants in the 2-qubit matrix,

$$\left[\begin{array}{c|c} \left(\begin{array}{c|c} (\frac{1}{2}) & (\frac{1}{2}) \\ \hline (\frac{1}{2}) & -\frac{1}{2} \end{array}\right) & \left(\begin{array}{c|c} (\frac{1}{2}) & (\frac{1}{2}) \\ \hline (\frac{1}{2}) & -\frac{1}{2} \end{array}\right) \\ \hline \left(\begin{array}{c|c} (\frac{1}{2}) & (\frac{1}{2}) \\ \hline (\frac{1}{2}) & -\frac{1}{2} \end{array}\right) & \begin{array}{c|c} (-\frac{1}{2}) & (-\frac{1}{2}) \\ \hline (-\frac{1}{2}) & \frac{1}{2} \end{array} \end{array}\right]$$

The only difference between the values in the 2-qubit matrix and those in the 1-qubit matrices is a factor of $1/\sqrt{2}$. Thus, we can recursively define the Hadamard operator as follows:

$$H^n = H \otimes H^{n-1} = \begin{bmatrix} C_1 H^{n-1} & C_1 H^{n-1} \\ C_1 H^{n-1} & C_2 H^{n-1} \end{bmatrix}$$

where $C_1 = 1/\sqrt{2}$ and $C_2 = -1/\sqrt{2}$. Other operators constructed via the tensor product can be defined recursively in a similar fashion.

Since three of the four blocks in an n-qubit Hadamard operator are equal, significant redundancy is present. The interleaved variable ordering property allows a QuIDD to explicitly represent only two distinct blocks rather than four, as shown in Figure 7.3. Later sections will demonstrate that compression of equivalent blocks using QuIDDs offers major performance improvements for many of the operators used frequently in quantum computation.

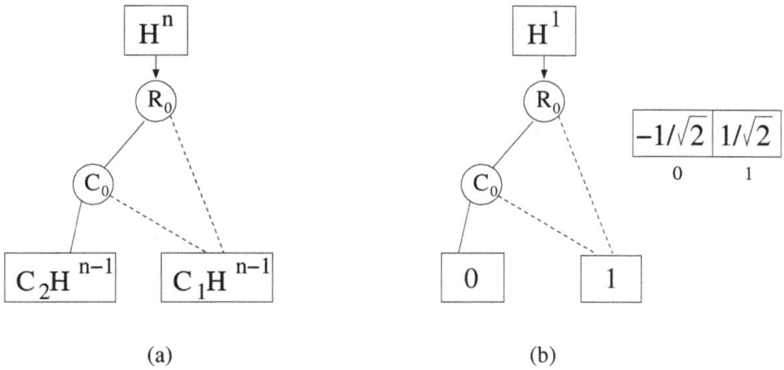

Fig. 7.3. (a) n-qubit Hadamard QuIDD depicted next to (b) 1-qubit Hadamard QuIDD. Notice that they are isomorphic except at the terminal nodes.

Operations on QuIDDs

Next, we describe the implementation of various linear-algebraic operations using QuIDDs.

Tensor Product. Most operations defined for ADDs also work on QuIDDs with minor modifications. The tensor (Kronecker) product has been described by Clarke et al. for MTBDDs representing various arithmetic matrices [25]. Our algorithm for the tensor product of QuIDDs is based on the **Apply** operation and is similar to that of [25].

Recall from Section 3.1 that the tensor product $A \otimes B$ requires multiplying each element of A by the entire matrix B. Rows (columns) of the tensor product matrix are component-wise products of rows (columns) of the argument matrices. Therefore, it is straightforward to implement the tensor product operation on QuIDDs using the **Apply** function with an argument that directs **Apply** to multiply when it reaches the terminals of both operands.

The main difficulty here lies in ensuring that each terminal of A is multiplied by *all* the terminals of B. From the definition of the standard recursive **Apply** routine, we know that variables which precede other variables in the ordering are expanded first [20, 25]. Therefore, we must first shift all variables in B in the current order *after* all of the variables in A, prior to the call to **Apply**. After this shift is performed, the **Apply** routine will then produce the desired behavior. **Apply** starts out with $A * B$ and expands A alone until $A_{terminal} * B$ is reached for each terminal in A. Once a terminal of A is reached, B is fully expanded, implying that each terminal of A is multiplied by all of B.

The size of the resulting QuIDD $A \otimes B$ and the runtime for generating it from two operands A and B of sizes $|A|$ and $|B|$ (in number of nodes) is $O(|A||B|)$ because the tensor product simply involves a variable shift of complexity $O(|B|)$, followed by a call to **Apply**, which Bryant showed to have time and memory complexity $O(|A||B|)$ [20].

Matrix Multiplication. Matrix multiplication can be implemented very efficiently by using **Apply** to implement the dot product operation. This follows from the observation that multiplication is a series of dot products between the rows of one operand and the columns of the other operand. In particular, given matrices A and B with elements a_{ij} and b_{ij}, their product $C = AB$ can be computed element-wise by $c_{ij} = \Sigma_{j=1}^{n} a_{ij} b_{ji}$.

Matrix multiplication for QuIDDs is an extension of the **Apply** function that implements the dot product. One call to **Apply** will not suffice because the dot product requires *two* binary operations to be performed, namely addition and multiplication. To implement this, we simply use the matrix multiplication algorithm defined by Bahar et al. for ADDs [6] but modified to support the QuIDD properties. The algorithm essentially makes two calls to **Apply**, one for multiplication and the other for addition.

Another important issue in efficient matrix multiplication is compression. To avoid the same problem that MATLAB encounters with its "packed" representation, ADDs do not require decompression during matrix multiplication. Bahar et al. [6] addressed this by tracking the number i of "skipped" variables between the parent node and its child node in each recursive call. To illustrate, suppose that $Var(v_f) = x_2$ and $Var(T(v_f)) = x_5$. In this situation, $i = 5 - 2 = 3$. A factor of 2^i is multiplied by the terminal-terminal product that is reached at the end of a recursive traversal [6].

The pseudo-code presented for this algorithm suggests time-complexity $O((|A||B|)^2)$, where A and B are two ADDs [6]. As with all algorithms based on **Apply**, the size of the resulting ADD is on the order of the time complexity, that is $O((|A||B|)^2)$. For QuIDDs, we use a modified form of this algorithm to multiply operators by the state vector, meaning that $|A|$ and $|B|$ will be the sizes in nodes of a QuIDD matrix and QuIDD state vector, respectively. If either a or b or both are exponential in the number of qubits in the circuit, the QuIDD approach will have exponential time and memory complexity. However, in Section 7.2 we prove that many of the operators which arise in quantum computing have QuIDD representations that are polynomial in the number of qubits.

Two important modifications must be made to the ADD matrix multiply algorithm in order to adapt it for QuIDDs. To satisfy QuIDD properties 1 and 2, the algorithm must treat the terminals as indices into an array rather than the actual values to be multiplied and added. Also, variable ordering must be accounted for when multiplying a matrix by a vector. A QuIDD matrix is composed of interleaved row and column variables, whereas a QuIDD vector only depends on column variables. If the ADD algorithm is run as described above without modification, the resulting QuIDD vector will be composed of row instead of column variables. The structure will be correct, but the dependence on row variables prevents the QuIDD vector from being used in future multiplications. Thus, we introduce a simple extension which transposes the row variables in the new QuIDD vector to the corresponding column variables. In other words, for each R_i variable that exists in the QuIDD vector's support, we map that variable to C_i.

Other Operations. Matrix addition is easily implemented by calling **Apply** with *op* defined to be addition. Unlike the tensor product, no special variable order shifting is required for matrix addition. Another interesting operation which is nearly identical to matrix addition is element-wise multiplication $c_{ij} = a_{ij}b_{ij}$. Unlike the dot product, this operation involves only products and no summation. This algorithm is implemented just like matrix addition except that *op* is defined to be multiplication rather than addition. Since matrix addition, element-wise multiplication, and their scalar counterparts are nothing more than calls to **Apply**, the runtime complexity of each operation is $O(|A||B|)$. Likewise, the resulting QuIDD has memory complexity $O(|A||B|)$ [20].

Another relevant operation which can be performed on QuIDDs is the transpose. It is perhaps the simplest QuIDD operation because it is accomplished merely by swapping the row and column variables. The transpose is easily extended to the complex conjugate transpose by first performing the transpose of a QuIDD and then conjugating its terminal values. The runtime and memory complexity of these operations is $O(a)$ where a is the size in nodes of the QuIDD undergoing a transpose.

Using the transpose, the inner product can be defined for QuIDDs. The inner product of two QuIDD vectors, e.g., $\langle A|B\rangle$, is computed by matrix multiplying the transpose of A with B. Since matrix multiplication is involved, the runtime and memory complexity of the inner product is $O((|A||B|)^2)$. Our QuIDD-based simulator QuIDDPro supports matrix multiplication, the tensor product, measurement, matrix addition, element-wise multiplication, scalar operations, the transpose, the complex conjugate transpose, and the inner product.

Measurement This can be defined for QuIDDs using a combination of operations. After measurement, the state vector is described by

$$\frac{M_m|\psi\rangle}{\sqrt{\langle\psi|M_m^\dagger M_m|\psi\rangle}}.$$

M_m is the measurement operator and can be represented by a QuIDD matrix, and the state vector $|\psi\rangle$ can be represented by a QuIDD vector. The numerator involves a QuIDD matrix multiplication. In the denominator, $M_m^\dagger$ is the complex conjugate transpose of M_m, which is also defined for QuIDDs. $M_m^\dagger M_m$ and $M_m^\dagger M_m|\psi\rangle$ are matrix multiplications. $\langle\psi|M_m^\dagger M_m|\psi\rangle$ is an inner product which produces a QuIDD with a single terminal node. Taking the square root of the value in this terminal node is straightforward. To complete the measurement, scalar division is performed with the QuIDD in the numerator and the single terminal QuIDD in the denominator as operands.

There are two ways two compute the measurement result. The first way is inefficient and involves computing the above formula explicitly. Performing the matrix multiplication in the numerator has runtime and memory complexity $O((|A||B|)^2)$. The scalar division of the numerator by the denominator also has the same runtime and memory complexity since the denominator is a QuIDD with a single terminal node. However, computing the denominator will have runtime and memory complexity $O(|A|^{16}|B|^6)$ due to the matrix-vector multiplications and inner product. A more efficient method is to multiply the measurement operator as before, but instead of computing the denominator, two calls to **Apply** are made. The first call uses **Apply** to determine the norm of the state vector. The second call divides each terminal value by the norm. The dominating complexity of all these operation is due to matrix multiplication, resulting in a runtime and memory complexity of $O((|A||B|)^2)$ for measurement.

Numerical Aspects

A software implementation of QuIDDs requires high-precision complex arithmetic because values like $(1+i)/\sqrt{2}$ and $1/2^n$ appear frequently in calculations. In principle, ADDs can support terminals of any numerical type. However, practical considerations motivate an implementation approach where terminal values are stored in an external array and accessed using indices of terminals. This approach makes it easy to use a variety of available numerical representations and simplifies experimentation.

Standard IEEE double-precision floating-point types can be used for small examples, but do not provide sufficient precision for simulating larger circuits. Round-off errors can significantly affect the structure of a QuIDD by merging terminals that are only slightly different, or not merging terminals whose values should be equal but differ by a small computational error ϵ. The use of approximate comparisons with an absolute additive ε works in certain cases but does not scale well, particularly for creating an equal superposition of states (a standard operation in quantum circuits). In an equal superposition, a circuit with n qubits contains the terminal value $\frac{1}{2^{n/2}}$ in the state vector. With the IEEE double-precision floating-point type, this value will be rounded to 0 at $n = 2048$, preventing the use of epsilons for approximate comparison past $n = 2048$. Furthermore, a static value for ε will not work well for circuits of very different sizes. For example, $\varepsilon = 10^{-6}$ may work well for $n = 35$, but not for $n = 40$ because at $n = 40$, all values may be smaller than 10^{-6}. To ensure adequate precision, our simulation software QuIDDPro uses an arbitrary-precision floating-point type from the GMP library [35] in conjunction with the C++ complex template. Precision is then limited only by the available amount of memory. Additionally, rather than using an additive ε, QuIDDPro uses a *relative* ϵ. This simply means that two values are considered equal if their difference is less than an ϵ fraction of the larger value.

Symbolic Arithmetic

Numerical accuracy of QuIDD operations can be increased through the use of symbolic number representations and related arithmetic operations. Such number representations can be built using conventional or high-precision integer types. For example, rational numbers $\frac{p}{q}$ can be represented as pairs of integers and numerically manipulated in this form. Complex rationals can be represented using four integers $\frac{p_{re}}{q_{re}} + \frac{p_{im}}{q_{im}}i$. Object-oriented encapsulation of such number representations in C++ classes with operator overloading facilitates easy replacement and experimentation.

Rationals can be further extended to represent some commonly occuring algebraic irrational numbers, such as *quadratic surds*. Consider the following rules for multiplication and division, where R may take on values such as $\sqrt{-1}$ or $\sqrt{2}$:

$$\begin{aligned}(a + b\sqrt{R})(c + d\sqrt{R}) &= ac + Rbd + (ad + bc)\sqrt{R} \\ (a + b\sqrt{R})/(c + d\sqrt{R}) &= (a + b\sqrt{R})(\tfrac{c}{c^2 - Rd^2} - \tfrac{d}{c^2 - Rd^2}\sqrt{R})\end{aligned}$$

Arithmetic operations involving the a, b, c, and d terms may be performed explicitly, while the root terms are represented symbolically relying on rules and reductions. Such representations improve precision for some numbers commonly occuring during quantum simulations, but cannot represent all relevant complex numbers. Additionally, to be useful in the hash tables that implement DD reduction rules, symbolic representations require non-trivial hashing functions. Linear or binary search may be used instead of hashing, but at increased (multiplicative) asymptotic cost of $O(n)$ or $O(\log n)$, where n is the number of different terminal values.

The potential reductions in DD size offered by symbolic representations of edge values and terminal values can be significant because less accurate representations suffer from slight numerical errors that can mask DD reductions by creating logically redundant nodes. Despite the obvious benefits of such reductions, symbolic representations require very careful implementations to avoid pitfalls, such as overlooked algebraic identities. Developing better symbolic representations for more efficient quantum simulation is an open problem applicable to most DD techniques.

7.2 Scalability of QuIDD-based Simulation

In this section, we prove that the QuIDD data structure can represent a large class of state vectors and operators using an amount of memory that grows linearly in the number of qubits, rather than exponentially. Further, we show that the basic QuIDD operations, matrix multiplication, the tensor product, and measurement, have both runtime and memory that is linear in the number of qubits for the same class of state vectors and operators. We also analyze the complexity of simulating Grover's algorithm using QuIDDs.

Complexity of QuIDD Operations

The key to analyzing the runtime and memory complexity of QuIDD-based simulation lies in the mechanics of the tensor product. In the following analysis, the size of a QuIDD is represented by the number of nodes rather than actual memory consumption. Since the amount of memory used by a single QuIDD node is a constant, size in nodes is relevant for asymptotic complexity arguments. Actual memory usage in megabytes of QuIDD simulations is reported in Section 7.3.

Figure 7.4 illustrates the general form of a tensor product between two QuIDDs A and B. $In(A)$ represents the internal nodes of A, while $Term(A)$ denotes the terminal nodes. The notation for B is similar. $In(A)$ is the root subgraph of the tensor product result because of the interleaved variable ordering defined for QuIDDs and the variable shifting operation of the tensor product (see Section 7.1).

Suppose that A depends on the variables $R_0 \prec C_0 \prec \ldots \prec R_i \prec C_i$, and B depends on the variables $R_0 \prec C_0 \prec \ldots \prec R_j \prec C_j$. In performing $A \otimes B$, the variables on which B depends will be shifted to $R_{i+1} \prec C_{i+1} \prec \ldots \prec R_{k+i+1} \prec C_{k+i+1}$. The tensor product is then completed by calling **Apply**$(A, B, *)$. Due to the variable shift on B, Rule 1 of the **Apply** function (Section 2.2) will be used recursively after each comparison of a node from A with a node from B until the terminals of A are reached. Using Rule 1 for each of these comparisons implies that only nodes from A will be added to the result, explaining the presence of $In(A)$. Once the terminals of A are reached, Rule 2 of **Apply** will then be invoked since terminals are defined to appear last in the variable ordering. Using Rule 2 when the terminals of A are reached implies that all the internal nodes from B will be added in place of each terminal of A, causing x copies of $In(B)$ to appear in the result (recall that there are x terminals in A). When the terminals of B are reached, they are multiplied by the appropriate terminals of A. Specifically, the terminals of a copy of B will each be multiplied by the terminal of A that its $In(B)$ replaced. The same reasoning holds for QuIDD vectors which differ in that they depend only on R_i variables.

Figure 7.4 suggests that the size of a QuIDD constructed via the tensor product depends on the number of terminals in the operands. The more ter-

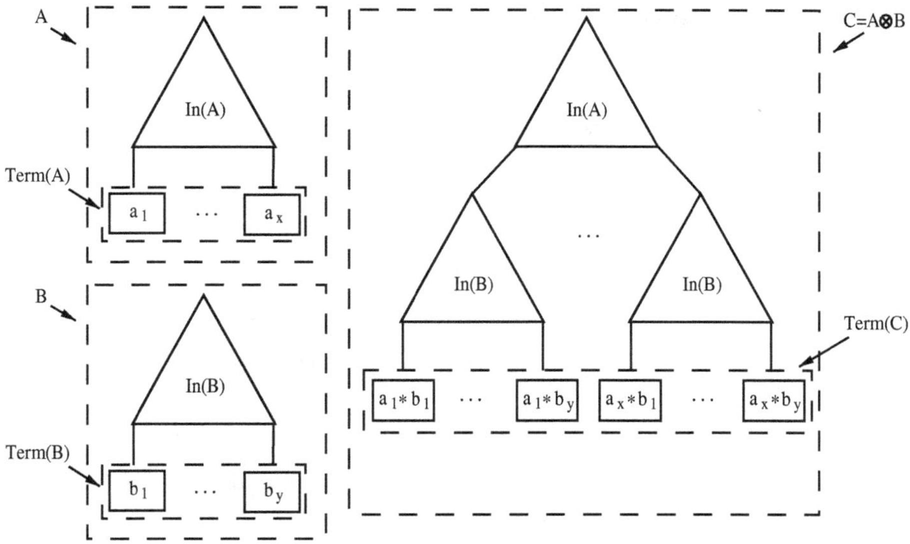

Fig. 7.4. General form of the tensor product C formed from QuIDDs A and B.

minals a left-hand tensor operand contains, the more copies of the right-hand tensor operand's internal nodes will be added to the result. More formally, consider the tensor product of a series of QuIDDs $\otimes_{i=1}^{n} Q_i = (\ldots((Q_1 \otimes Q_2) \otimes Q_3) \otimes \ldots \otimes Q_n)$. Note that the $\otimes$ operation is associative (thus parentheses do not affect the result), but it is not commutative. The number of nodes in this tensor product is described by the following lemma. Let $|In(A)|$ denote the number of internal nodes in A, while $|Term(A)|$ denotes the number of terminal nodes in A.

Lemma 7.1. *The tensor-product QuIDD* $\otimes_{i=1}^{n} Q_i$ *contains* $|In(Q_1)| + \Sigma_{i=2}^{n} |In(Q_i)||Term(\otimes_{j=1}^{i-1} Q_j)| + |Term(\otimes_{i=1}^{n} Q_i)|$ *nodes.*

Proof. This formula can be verified by induction. For the base case, $n = 1$, there is a single QuIDD Q_1. Putting this information into the formula eliminates the summation term, leaving $|In(Q_1)| + |Term(Q_1)|$ as the total number of nodes in Q_1. This is clearly correct since, by definition, a QuIDD is composed of its internal and terminal nodes. To complete the proof, we now show that if the formula is true for Q_n then it's true for Q_{n+1}. The inductive hypothesis for Q_n is $| \otimes_{i=1}^{n} Q_i| = |In(Q_1)| + \Sigma_{i=2}^{n} |In(Q_i)||Term(\otimes_{j=1}^{i-1} Q_j)| + |Term(\otimes_{i=1}^{n} Q_i)|$. For Q_{n+1} the number of nodes is

$$\begin{aligned} |(\otimes_{i=1}^{n} Q_i) \otimes Q_{n+1}| = &\; | \otimes_{i=1}^{n} Q_i| - |Term(\otimes_{i=1}^{n} Q_i)| \\ &+ |In(Q_{n+1})||Term(\otimes_{i=1}^{n} Q_i)| + |Term(\otimes_{i=1}^{n+1} Q_i)| \end{aligned}$$

Notice that the number of terminals in $\otimes_{i=1}^{n} Q_i$ is subtracted from the total number of nodes in $\otimes_{i=1}^{n} Q_i$ and multiplied by the number of internal nodes in

Q_{n+1}. The presence of these terms is due to Rule 2 of **Apply** which dictates that in the tensor product $(\otimes_{i=1}^{n} Q_i) \otimes Q_{n+1}$, the terminals of $\otimes_{i=1}^{n} Q_i$ are replaced by copies of Q_{n+1} where each copy's terminals are multiplied by a terminal from $\otimes_{i=1}^{n} Q_i$. The last term simply accounts for the total number of terminals in the tensor product. Substituting the inductive hypothesis made earlier for the term $|\otimes_{i=1}^{n} Q_i|$ produces

$$|In(Q_1)| + \Sigma_{i=2}^{n} |In(Q_i)||Term(\otimes_{j=1}^{i-1} Q_j)| + |Term(\otimes_{i=1}^{n} Q_i)| -$$
$$|Term(\otimes_{i=1}^{n} Q_i)| + |In(Q_{n+1})||Term(\otimes_{i=1}^{n} Q_i)| + |Term(\otimes_{i=1}^{n+1} Q_i)|$$
$$= |In(Q_1)| + \Sigma_{i=2}^{n+1} |In(Q_i)||Term(\otimes_{j=1}^{i-1} Q_j)| + |Term(\otimes_{i=1}^{n+1})|$$

Thus the number of nodes in Q_{n+1} satisfies the original formula we set out to prove for $n+1$, and the induction is complete. □

Lemma 7.1 suggests that if the number of terminals in $\otimes_{i=1} Q_i$ increases by a certain factor with each Q_i, then $\otimes_{i=1}^{n} Q_i$ must grow exponentially in n. If, however, the number of terminals stops changing, then $\otimes_{i=1}^{n} Q_i$ must grow linearly in n. Thus, the growth depends on matrix entries because terminals of $A \otimes B$ are products of terminal values of A by terminal values of B and repeated products are merged. If all QuIDDs Q_i have terminal values from the same set Γ, the product's terminal values are products of elements from Γ.

Definition 7.2. *Consider finite non-empty sets of complex numbers Γ_1 and Γ_2, and define their* all-pairs product *as $\{xy \mid x \in \Gamma_1,\ y \in \Gamma_2\}$. This operation is associative, so the set Γ^n of* all n-element products *is well-defined for $n > 0$. We then call a finite non-empty set $\Gamma \subset \mathbb{C}$* persistent *if and only if the size of Γ^n is constant for all $n > 0$.*

For example, the set $\Gamma = \{c, -c\}$ is persistent for any c because $\Gamma^n = \{c^n, -c^n\}$. In general, any set closed under multiplication is persistent, but that is not a necessary condition. In particular, for $c \neq 0$, the persistence of Γ is equivalent to the persistence of $c\Gamma$. Another observation is that Γ is persistent if and only if $\Gamma \cup \{0\}$ is persistent. An important example of a persistent set is the set consisting of 0 and all n-th degree roots of unity $\mathbb{U}_n = \{e^{2\pi ik/n} | k = 0..n-1\}$, for some n. Since roots of unity form a group, they are closed under multiplication and form a persistent set.

The following sequence of lemmas leads to a complete characterization of persistent sets from Definition 7.2 [99]. This definition considers finite non-empty sets of complex numbers Γ_1 and Γ_2, and denotes their *all-pairs product* as $\{xy \mid x \in \Gamma_1,\ y \in \Gamma_2\}$. One can verify that this operation is associative, and therefore the set Γ^n of *all n-element products* is well-defined for $n > 0$. We then call a finite non-empty set $\Gamma \subset \mathbb{C}$ *persistent* if and only if the size of Γ^n is constant for all $n > 0$. We start by observing that adding 0 to, or removing 0 from, a set does not affect its persistence.

Lemma 7.3. All elements of a persistent set Γ that does not contain 0 must have the same magnitude.

Proof. For Γ to be persistent, the set of magnitudes of elements from Γ must also be persistent. Therefore, it suffices to show that each persistent set of positive real numbers contains no more than one element. Assume, by way of contradiction, that such a persistent set exists with at least two elements r and s. Then among the n-element products from Γ, we find all numbers of the form $r^{n-k}s^k$ for $k = 0..n$. If we order r and s so that $r < s$, then it becomes clear that the products are all different because $r^{n-k+1}s^{k-1} < r^{n-k}s^k$. □

Lemma 7.4. All persistent sets without 0 are of the form $c\Gamma'$, where $c \neq 0$ and Γ' is a finite persistent subset of the unit circle in the complex plane $\mathbb{C}$ that contains 1 and is closed under multiplication. Conversely, for all such sets Γ' and $c \neq 0$, $c\Gamma'$ is persistent.

Proof. Take a persistent set Γ that does not contain 0, pick an element $z \in \Gamma$, and define $\Gamma' = \Gamma/z$, which is persistent by construction. Γ' is a subset of the unit circle because all numbers in Γ have the same magnitude. Due to the fact that $z/z = 1 \in \Gamma'$, the set of n-element products contains every element of Γ'. Should the product of two elements of Γ' fall outside the set, Γ' cannot be persistent. □

Lemma 7.5. A finite persistent subset Γ' of the unit circle that contains 1 and is closed under multiplication must be of the form $\mathbb{U}_n$ (roots of unity of degree n).

Proof. If $\Gamma' = \{1\}$, then $n = 1$, and we are done. Otherwise consider an arbitrary element $z \neq 1$ of Γ' and observe that all powers of z must also be in Γ'. Since Γ' is finite, $z^m = z^k$ for some $m \neq k$, hence $z^{m-k} = 1$, and z is a root of unity. Therefore, Γ' is closed under inversion, and forms a group. It follows from group theory, that a finite subgroup of $\mathbb{C}$ is necessarily of the form $\mathbb{U}_n$ for some n. □

Theorem 7.6. Persistent sets are either of the form $c\mathbb{U}_n$ for some $c \neq 0$ and n, or $\{0\} \cup c\mathbb{U}_n$.

The importance of persistent sets is underlined by the following theorem.

Theorem 7.7. *Given a persistent set Γ and a constant C, consider n QuIDDs with at most C nodes each and terminal values from Γ. The tensor product of those QuIDDs has $O(n)$ nodes and can be computed in $O(n)$ time.*

Proof. The first and the last terms of the formula in Lemma 7.1 are bounded by C and $|\Gamma|$ respectively. As the sizes of terminal sets in the middle term are bounded by $|\Gamma|$, the middle term is bounded by $|\Gamma|\sum_{i=2}^{n}|In(Q_i)| < |\Gamma|c$ since each $|In(Q_i)|$ is a constant. The tensor product operation $A \otimes B$ for QuIDDs involves a shift of variables on B followed by **Apply**$(A, B, *)$. If

B is a QuIDD representing n qubits, then B depends on $O(n)$ variables. (More precisely, B depends on exactly $2n$ variables if it is a matrix QuIDD, and n variables if it is a vector QuIDD.) This implies that the runtime of the variable shift is $O(n)$. Bryant proved that the asymptotic runtime and memory complexity of $\mathbf{Apply}(A, B, binary\_op)$ is $O(|A||B|)$ [20]. Lemma 7.1 and the fact that we are considering QuIDDs with at most C nodes and terminals from a persistent set Γ imply that $|A| = O(n)$ and $|B| = O(1)$. Thus, $\mathbf{Apply}(A, B, *)$ has asymptotic runtime and memory complexity $O(n)$, leading to an overall asymptotic runtime and memory complexity of $O(n)$ for computing $\otimes_{i=1}^{n} Q_i$. □

Importantly, the terminal values do not need to form a persistent set themselves for the theorem to hold. If they are *contained* in a persistent set, then the sets of all possible m-element products, i.e., $m \leq n$ for all n-element products in a set Γ, eventually stabilize in the sense that their sizes do not exceed that of Γ. However, this is only true for a fixed m rather than for the sets of products of m elements and fewer.

For QuIDDs A and B, the matrix-matrix and matrix-vector product computations are not as sensitive to terminal values, but depend on the sizes of the QuIDDs. Indeed, the memory and time complexity of this operation is $O((|A||B|)^2)$ [6].

Theorem 7.8. *Consider measuring an n-qubit QuIDD state vector $|\psi\rangle$ using a QuIDD measurement operator M, where both $|\psi\rangle$ and M are constructed via the tensor product of an arbitrary sequence of $O(1)$-sized QuIDD vectors and matrices, respectively. If the terminal node values of the QuIDDs are in a persistent set Γ, then the runtime and memory complexity of measuring the QuIDD state vector is $O(n^4)$. Note that this worst-case bound is rarely reached in practice, as demonstrated later.*

Proof. We have previously shown that runtime and memory complexity for measuring a state vector QuIDD are $O((|A||B|)^2)$, where $|A|$ and $|B|$ are the sizes in nodes of the measurement operator QuIDD and state vector QuIDD, respectively. From Theorem 7.7, the asymptotic memory complexity of both $|A|$ and $|B|$ is $O(n)$, leading to an overall runtime and memory complexity of $O(n^4)$. □

The class of QuIDDs described by Theorem 7.7 and its corollaries, with terminals taken from the set $\{0\} \cup c\mathbb{U}$, encompass a large number of practical quantum state vectors and operators. These include, but are not limited to, any equal superposition of n qubits, any sequence of n qubits in the computational basis states, n-qubit Pauli matrices, and n-qubit Hadamard matrices. The above results suggest a polynomial-sized QuIDD representation of any quantum circuit on n qubits in terms of such gates if the number of gates is limited by a constant. In other words, the above sufficient conditions apply if the depth (length) of the circuit is limited by a constant. Our simulation technique may use polynomial memory and runtime in other circumstances as well, as shown next.

Simulating Grover's Algorithm

To investigate the power of the QuIDD representation, we used QuIDDPro to simulate Grover's algorithm [38], one of the few major quantum algorithms that have been developed to date. The high-level functionality of the algorithm is described in Subsection 4.4. Here we provide more detailed steps which will be important when proving aspects of QuIDD complexity for simulating the algorithm. These steps may be summarized as follows, with N denoting the number of elements in the database.

Grover's Algorithm

Step 1. Initialize $n = \lceil \log_2 N \rceil$ qubits to $|0\rangle$ and the *oracle qubit* to $|1\rangle$.

Step 2. Apply the Hadamard gate H to all qubits to put them into a uniform superposition of basis states.

Step 3. Apply the oracle operation, which can be implemented as a series of one or more CNOT gates representing the search criteria. The inputs to the oracle circuit feed into the control portions of the CNOT gates, while the oracle qubit is the target qubit for all of the CNOT gates. In this way, if the input to this circuit satisfies the search criteria, the state of the oracle qubit is flipped. For a superposition of inputs, those input basis states that satisfy the search criteria flip the oracle qubit in the composite state-space. The oracle circuit uses ancillary qubits as its workspace, reversibly returning them to their original states (shown as $|0\rangle$ in Fig 4.6). These ancillary qubits will not be operated on by any other step in the algorithm.

Step 4. Apply the H gate to all qubits except the oracle qubit.

Step 5. Apply the phase-shift gate on all qubits except the oracle qubit. This gate negates the probability amplitude of the $|000\ldots0\rangle$ basis state, leaving that of the others unaffected. It can be realized using a combination of X, H and C^{n-1}-NOT gates as shown. A decomposition of the C^{n-1}-NOT into elementary gates is given in [7].

Step 6. Apply the H gate to all qubits except the oracle qubit.

Step 7. Repeat Steps 3-6 (a single Grover iteration) R times, where $R = \lfloor \frac{\pi}{4}\sqrt{\frac{N}{M}} \rfloor$ and M is the number of keys matching the search criteria [17].

Step 8. Apply the H gate to the oracle qubit in the last iteration. Measure the first n qubits to obtain the index of the matching key with high probability.

Using explicit vectors and matrices to simulate the above procedure would incur memory and runtime complexities of $\Omega(2^n)$. However, this is not necessarily the case when using QuIDDs. To show this, we present a step-by-step complexity analysis for a QuIDD-based simulation of the procedure.

Steps 1-2. Theorem 7.7 implies that the memory and runtime complexity of Step 1 is $O(n)$ because the initial state vector only contains elements in $c\mathbb{U}_k \cup \{0\}$ and is constructed via the tensor product. Step 2 is simply a matrix multiplication of an n-qubit Hadamard matrix with the state vector constructed in Step 1. The Hadamard matrix has memory complexity $O(n)$ by Theorem 7.7. Since the state vector also has memory complexity $O(n)$, further matrix-vector multiplications in Step 2 each have $O(n^4)$ memory and runtime complexity because computing the product of two QuIDDs A and B takes $O((|A||B|)^2)$ time and memory [6]. This upper bound can be trivially tightened, however. The function of Steps 1 and 2 is to put the qubits into an equal superposition. For the n data qubits, this produces a QuIDD with $O(1)$ nodes because an n-qubit state vector representing an equal superposition has only one distinct element, namely $\frac{1}{2^{n/2}}$. Also, applying a Hadamard gate to the single oracle qubit results in a QuIDD with $O(1)$ nodes because in the worst-case, the size of a 1-qubit QuIDD is clearly a constant. Since the tensor product is based on the **Apply** algorithm, the result of tensoring the QuIDD representing the data qubits in an equal superposition with the QuIDD for the oracle qubit is a QuIDD containing $O(1)$ nodes.

Steps 3-6. In Step 3, the state vector is multiplied by the oracle matrix. Again, the complexity of multiplying two arbitrary QuIDDs A and B is $O((|A||B|)^2)$. The size of the state vector in Step 3 is $O(1)$. If the size of the oracle is $|A|$, then the memory and runtime complexity of Step 3 is $O(|A|^2)$. Similarly, Steps 4, 5 and 6 will have polynomial memory and runtime complexity in terms of $|A|$ and n. As noted in Step 5, the phase-shift operator can be decomposed into the tensor product of single qubit matrices, giving it memory complexity $O(n)$. Thus we arrive at the $O(|A|^{16}n^{14})$ worst-case upper bound for the memory and runtime complexity of the simulation at Step 6. Judging from our empirical data, this bound is typically very pessimistic.

Lemma 7.9. *The memory and runtime complexity of a single Grover iteration in a QuIDD-based simulation is* $O(|A|^{16}n^{14})$.

Proof. Steps 3-6 make up a single Grover iteration. Since the memory and runtime complexity of a QuIDD-based simulation after completing Step 6 is $O(|A|^{16}n^{14})$, the memory and runtime complexity of a single Grover iteration is $O(|A|^{16}n^{14})$. □

Step 7. This does not involve a quantum operator, but rather it repeats a Grover iteration $R = \lfloor \frac{\pi}{4}\sqrt{\frac{N}{M}} \rfloor$ times. As a result, Step 7 induces an exponential runtime for the simulation, since the number of Grover iterations is a function of $N = 2^n$. This is acceptable, because an actual quantum computer would also require exponentially many Grover iterations in order to measure one of the matching keys with a high probability [17]. Ultimately, this is the reason why Grover's algorithm only offers a *quadratic* and not an exponential speedup over classical search. Since Lemma 7.9 shows that the memory and runtime complexity of a single Grover iteration is polynomial in the size of the oracle QuIDD, one might guess that the memory complexity of Step 7 is exponential like the runtime. However, it turns out that the size of the state vector does not change from iteration to iteration, as shown below.

Lemma 7.10. *The number of internal nodes of the state vector QuIDD at the end of any Grover iteration i is equal to the number of internal nodes of the state vector QuIDD at the end of Grover iteration $i+1$.*

Proof. Each Grover iteration increases the probability of the states representing matching keys, while simultaneously decreasing the probability of the states representing non-matching keys. Therefore, at the end of the first iteration, the state vector QuIDD will have a single terminal node for all the states representing matching keys and one other terminal node, with a lower value, for the states representing non-matching keys (there may be two such terminal nodes for non-matching keys, depending on machine precision). The number of internal nodes of the state vector QuIDD cannot be different at the end of subsequent Grover iterations because a Grover iteration does not change the pattern of probability amplitudes, only their values. In other words, the same matching states always point to a terminal node whose value becomes closer to 1 after each iteration, while the same non-matching states always point to one or more terminal nodes whose values becomes closer to 0. □

Lemma 7.11. *The total number of nodes in the state vector QuIDD at the end of any Grover iteration i is equal to the total number of nodes in the state vector QuIDD at the end of Grover iteration $i + 1$.*

Proof. In proving Lemma 7.10, we showed that the only change in the state vector QuIDD from iteration to iteration is in the values in the terminal nodes (not in the number of terminal nodes). Therefore, the number of nodes in the state vector QuIDD is the same at the end of all Grover iterations. □

Corollary 7.12. *In a QuIDD-based simulation, the runtime and memory complexity of any Grover iteration i is equal to the runtime and memory complexity of Grover iteration $i + 1$.*

Proof. Each Grover iteration is a series of matrix multiplications between the state vector QuIDD and several operator QuIDDs (Steps 3-6). The work of Bahar et al. shows that matrix multiplication with ADDs has runtime and memory complexity that is determined solely by the number of nodes in the operands (see Section 7.1) [6]. Since the total number of nodes in the state vector QuIDD is always the same at the end of every Grover iteration, the runtime and memory complexity of every Grover iteration is the same. □

Lemmas 7.10 and 7.11 imply that Step 7 does not necessarily induce memory complexity that is exponential in the number of qubits. This important fact is captured in the following theorem.

Theorem 7.13. *The memory complexity of simulating Grover's algorithm using QuIDDs is polynomial in the size of the oracle QuIDD and the number of qubits.*

Proof. The runtime and memory complexity of a single Grover iteration is $O(|A|^{16}n^{14})$ (Lemma 7.9), which includes the initialization costs of Steps 1 and 2. Also, the structure of the state vector QuIDD does not change from one Grover iteration to the next (Lemmas 7.10 and 7.11). Thus, the overall memory complexity of simulating Grover's algorithm with QuIDDs is $O(|A|^{16}n^{14})$,

Circuit size n	Hadamard gates Initial	Repeated	Phase-shift gates	Oracles 1	2
20	80	83	21	99	108
30	120	123	31	149	168
40	160	163	41	199	228
50	200	203	51	249	288
60	240	243	61	299	348
70	280	283	71	349	408
80	320	323	81	399	468
90	360	363	91	449	528
100	400	403	101	499	588

Table 7.1. Size of QuIDDs in number of nodes for Grover's algorithm.

where $|A|$ is the number of nodes in the oracle QuIDD and n is the number of qubits. □

Note that Theorem 7.13 does not account for the resources required to construct the oracle QuIDD. While any polynomial-time quantum computation can be simulated in polynomial space, the commonly-used linear-algebraic simulation requires $\Omega(2^n)$ space. Also note that the case of an oracle searching for a unique solution (originally considered by Grover) implies that $|A| = n$. Here, most of the searching will be done while constructing the QuIDD of the oracle, which is an entirely classical operation.

As demonstrated experimentally in Section 7.3, for some oracles, simulating Grover's algorithm with QuIDDs has memory complexity $\Theta(n)$. Furthermore, simulation using QuIDDs has worst-case runtime complexity $O(R|A|^{16}n^{14})$, where R is the number of Grover iterations as defined earlier. If $|A|$ grows polynomially with n, this runtime complexity is the same as that of an ideal quantum computer, up to a polynomial factor.

7.3 Empirical Validation

We now present empirical results obtained with our QuIDD-based simulator.

Simulating Grover's Algorithm

First, we construct the QuIDD representations of Hadamard operators by incrementally tensoring together one-qubit versions of their matrices $n-1$ times to get n-qubit versions. All other QuIDD operators are constructed similarly. Table 7.1 shows the size (in nodes) of the QuIDDs for n-qubits, where n ranges from 20 to 100. We observe that memory usage grows linearly in n, and as a result QuIDD-based simulations of Grover's algorithm are not memory-limited even at 100 qubits. This is consistent with Theorem 7.7.

Oracle 1: Runtime (s)				
n	Oct	MAT	B++	QP
10	80.6	6.64	0.15	0.33
11	2.65e2	22.5	0.48	0.54
12	8.36e2	74.2	1.49	0.83
13	2.75e3	2.55e2	4.70	1.30
14	1.03e4	1.06e3	14.6	2.01
15	4.82e4	6.76e3	44.7	3.09
16	> 24hrs	> 24hrs	1.35e2	4.79
17	> 24hrs	> 24hrs	4.09e2	7.36
18	> 24hrs	> 24hrs	1.23e3	11.3
19	> 24hrs	> 24hrs	3.67e3	17.1
20	> 24hrs	> 24hrs	1.09e4	26.2
21	> 24hrs	> 24hrs	3.26e4	39.7
22	> 24hrs	> 24hrs	> 24hrs	60.5
23	> 24hrs	> 24hrs	> 24hrs	92.7
24	> 24hrs	> 24hrs	> 24hrs	1.40e2
25	> 24hrs	> 24hrs	> 24hrs	2.08e2
26	> 24hrs	> 24hrs	> 24hrs	3.12e2
27	> 24hrs	> 24hrs	> 24hrs	4.72e2
28	> 24hrs	> 24hrs	> 24hrs	7.07e2
29	> 24hrs	> 24hrs	> 24hrs	1.08e3
30	> 24hrs	> 24hrs	> 24hrs	1.57e3
31	> 24hrs	> 24hrs	> 24hrs	2.35e3
32	> 24hrs	> 24hrs	> 24hrs	3.53e3
33	> 24hrs	> 24hrs	> 24hrs	5.23e3
34	> 24hrs	> 24hrs	> 24hrs	7.90e3
35	> 24hrs	> 24hrs	> 24hrs	1.15e4
36	> 24hrs	> 24hrs	> 24hrs	1.71e4
37	> 24hrs	> 24hrs	> 24hrs	2.57e4
38	> 24hrs	> 24hrs	> 24hrs	3.82e4
39	> 24hrs	> 24hrs	> 24hrs	5.64e4
40	> 24hrs	> 24hrs	> 24hrs	8.23e4

(a)

Oracle 1: Peak Memory Usage (MB)				
n	Oct	MAT	B++	QP
10	2.64e-2	1.05e-2	3.52e-2	9.38e-2
11	5.47e-2	2.07e-2	8.20e-2	0.121
12	0.105	4.12e-2	0.176	0.137
13	0.213	8.22e-2	0.309	0.137
14	0.426	0.164	0.559	0.137
15	0.837	0.328	1.06	0.137
16	1.74	0.656	2.06	0.145
17	3.34	1.31	4.06	0.172
18	4.59	2.62	8.06	0.172
19	13.4	5.24	16.1	0.172
20	27.8	10.5	32.1	0.172
21	55.6	NA	64.1	0.195
22	NA	NA	1.28e2	0.207
23	NA	NA	2.56e2	0.207
24	NA	NA	5.12e2	0.223
25	NA	NA	1.02e3	0.230
26	NA	NA	> 1.5GB	0.238
27	NA	NA	> 1.5GB	0.254
28	NA	NA	> 1.5GB	0.262
29	NA	NA	> 1.5GB	0.277
30	NA	NA	> 1.5GB	0.297
31	NA	NA	> 1.5GB	0.301
32	NA	NA	> 1.5GB	0.305
33	NA	NA	> 1.5GB	0.320
34	NA	NA	> 1.5GB	0.324
35	NA	NA	> 1.5GB	0.348
36	NA	NA	> 1.5GB	0.352
37	NA	NA	> 1.5GB	0.371
38	NA	NA	> 1.5GB	0.375
39	NA	NA	> 1.5GB	0.395
40	NA	NA	> 1.5GB	0.398

(b)

Table 7.2. Simulating Grover's algorithm with n qubits and Oracle 1 using Octave (Oct), MATLAB (MAT), Blitz++ (B++) and our simulator QuIDDPro (QP). *Notation:* $>$ 24hrs indicates that runtime exceeded 24 hours. $>$ 1.5GB indicates that memory usage exceeded 1.5GB. Simulation runs that exceed the memory cutoff may also exceed the time cutoff. NA indicates that after one week, memory usage was still growing.

With the operators constructed, simulation can proceed. Tables 7.2a and 7.2b show performance measurements for simulating Grover's algorithm with an oracle circuit (Oracle 1)that searches for one item out of 2^n. QuIDDPro achieves asymptotic memory savings compared to qubit-wise implementations

Oracle 2: Runtime (s)				
n	Oct	MAT	B++	QP
13	1.39e3	1.31e2	2.47	0.617
14	3.75e3	7.26e2	5.42	0.662
15	1.11e4	4.27e3	11.7	0.705
16	3.70e4	2.23e4	24.9	0.756
17	> 24hrs	> 24hrs	53.4	0.805
18	> 24hrs	> 24hrs	1.13e2	0.863
19	> 24hrs	> 24hrs	2.39e2	0.910
20	> 24hrs	> 24hrs	5.15e2	0.965
21	> 24hrs	> 24hrs	1.14e3	1.03
22	> 24hrs	> 24hrs	2.25e3	1.09
23	> 24hrs	> 24hrs	5.21e3	1.15
24	> 24hrs	> 24hrs	1.02e4	1.21
25	> 24hrs	> 24hrs	2.19e4	1.28
26	> 24hrs	> 24hrs	> 1.5GB	1.35
27	> 24hrs	> 24hrs	> 1.5GB	1.41
28	> 24hrs	> 24hrs	> 1.5GB	1.49
29	> 24hrs	> 24hrs	> 1.5GB	1.55
30	> 24hrs	> 24hrs	> 1.5GB	1.63
31	> 24hrs	> 24hrs	> 1.5GB	1.71
32	> 24hrs	> 24hrs	> 1.5GB	1.78
33	> 24hrs	> 24hrs	> 1.5GB	1.86
34	> 24hrs	> 24hrs	> 1.5GB	1.94
35	> 24hrs	> 24hrs	> 1.5GB	2.03
36	> 24hrs	> 24hrs	> 1.5GB	2.12
37	> 24hrs	> 24hrs	> 1.5GB	2.21
38	> 24hrs	> 24hrs	> 1.5GB	2.29
39	> 24hrs	> 24hrs	> 1.5GB	2.37
40	> 24hrs	> 24hrs	> 1.5GB	2.47

(a)

Oracle 2: Peak Memory Usage (MB)				
n	Oct	MAT	B++	QP
13	0.218	8.22e-2	0.252	0.137
14	0.436	0.164	0.563	0.141
15	0.873	0.328	1.06	0.145
16	1.74	0.656	2.06	0.172
17	3.34	1.31	4.06	0.176
18	4.59	2.62	8.06	0.180
19	13.4	5.24	16.1	0.180
20	27.8	10.5	32.1	0.195
21	55.6	NA	64.1	0.199
22	NA	NA	1.28e2	0.207
23	NA	NA	2.56e2	0.215
24	NA	NA	5.12e2	0.227
25	NA	NA	1.02e3	0.238
26	NA	NA	> 1.5GB	0.246
27	NA	NA	> 1.5GB	0.256
28	NA	NA	> 1.5GB	0.266
29	NA	NA	> 1.5GB	0.297
30	NA	NA	> 1.5GB	0.301
31	NA	NA	> 1.5GB	0.305
32	NA	NA	> 1.5GB	0.324
33	NA	NA	> 1.5GB	0.328
34	NA	NA	> 1.5GB	0.348
35	NA	NA	> 1.5GB	0.352
36	NA	NA	> 1.5GB	0.375
37	NA	NA	> 1.5GB	0.375
38	NA	NA	> 1.5GB	0.395
39	NA	NA	> 1.5GB	0.398
40	NA	NA	> 1.5GB	0.408

(b)

Table 7.3. Simulating Grover's algorithm with n qubits and Oracle 2 using Octave (Oct), MATLAB (MAT), Blitz++ (B++) and our simulator QuIDDPro (QP). *Notation:* > 24hrs indicates that runtime exceeded 24 hours. > 1.5GB indicates that memory usage exceeded 1.5GB. Simulation runs that exceed the memory cutoff may also exceed the time cutoff. NA indicates that after one week, memory usage was still growing.

(see Section 6.1) of Grover's algorithm using Blitz++, a high-performance numerical linear algebra library for C++ [94], MATLAB, and Octave, a mathematical package similar to MATLAB. The overall runtimes are still exponential in n because Grover's algorithm entails an exponential number of iterations, even on an actual quantum computer [17]. We also studied a "mod-1024" oracle circuit (Oracle 2) that searches for elements whose ten least significant bits are 1 (see Tables 7.3a and 7.3b). Results were produced on a 1.2GHz AMD Athlon with 1GB RAM running Linux. Memory usage for MATLAB and Octave is lower-bounded by the size of the state vector and phase-shift operator; Blitz++ and QuIDDPro memory usage is measured as the size of the entire program. Simulations using MATLAB and Octave past 15 qubits timed out at 24 hours.

Impact of Grover Iterations. To verify that the QuIDDPro simulation resulted in the exact number of Grover iterations required to generate the highest probability of measuring the items being sought as per the Boyer et al. formulation [17], we tracked the probabilities of these items as a function of the number of iterations. For this experiment, we used four different oracle circuits, each with 11, 12 and 13 qubit circuits. The first oracle is called "Oracle N" and represents an oracle in which all the data qubits act as controls to flip the oracle qubit (this oracle is equivalent to Oracle 1 in Table 7.2). The other oracle circuits are "Oracle N-1", "Oracle N-2", and "Oracle N-3", which all have the same structure as Oracle N minus 1, 2 and 3 controls, respectively. As described earlier, each removal of a control doubles the number of items being searched for in the database. For example, Oracle N-2 searches for 4 items in the data set because it recognizes the bit pattern 111...1*dd*.

Oracle	11 Qubits	12 Qubits	13 Qubits
N	25	35	50
$N-1$	17	25	35
$N-2$	12	17	25
$N-3$	8	12	17

Table 7.4. Number of Grover iterations at which Boyer et al. [17] predict the highest probability of measuring one of the items sought.

Table 7.4 shows the optimal number of iterations produced with the Boyer et al. formulation for all the instances tested. Figure 7.5 plots the probability of successfully finding any of the items sought against the number of Grover iterations. In the case of Oracle N, we plot the probability of measuring the single item being searched for. Similarly, for oracles N-1, N-2, and N-3, we plot the probability of measuring any one of the 2, 4, and 8 items being searched for, respectively. By comparing Table 7.4 with Figure 7.5, it can be verified that QuIDDPro uses the correct number of iterations at which measurement is most likely to produce items sought. Also notice that the probabilities, as a function of the number of iterations, follow sinusoidal curves whose minima and maxima match the predictions in Table 7.4. It is therefore important to terminate at the exact optimal number of iterations, not only from an efficiency standpoint but also to prevent the probability amplitudes of the items being sought from dropping down toward 0.

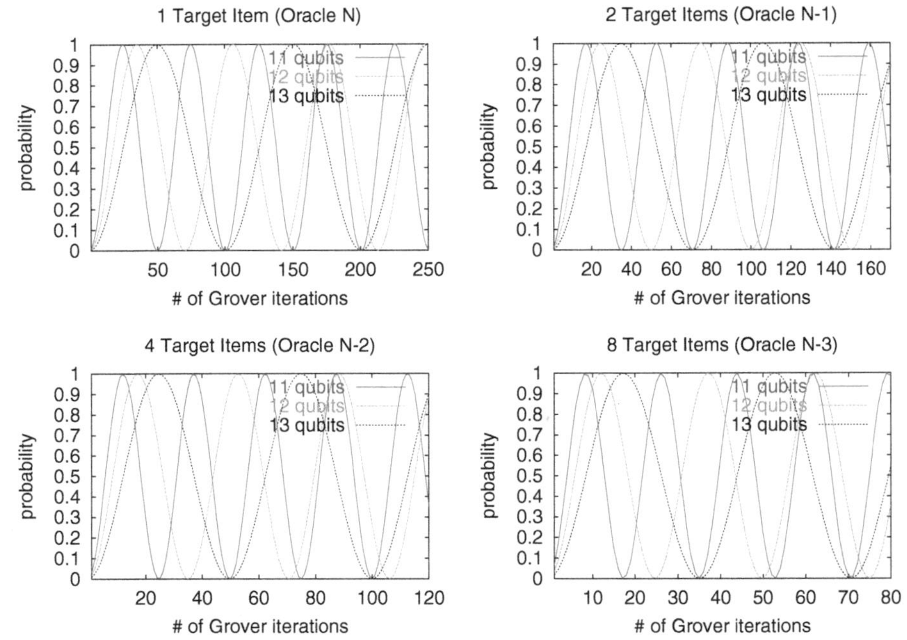

Fig. 7.5. Probability of successful search for 1, 2, 4 and 8 items as a function of the number of iterations after which the measurement is performed (11, 12 and 13 qubits).

7.4 Related Decision Diagrams

A natural extension to any decision diagram is the notion of edge values or weighted edges. By encoding more information into the graph structure, even more compact representations of vectors and matrices are possible [73]. However, edge values introduce new complexities into the **Apply**-based algorithms. If care is not taken, compression rates can suffer due to pitfalls such as loss of canonicity and an exponential increase in the number of paths that will be traversed in the worst case. While the QuIDD data structure as described so far does not use edge values, this section explores the benefits of such extensions to QuIDDs and to other decision diagram (DD) data structures.

Multiplicative Edge Values and QuIDDs

Although not implemented in QuIDDPro, it is useful to explore how QuIDDs could be extended directly with edge values. In particular, we will examine the use of multiplicative edge values with QuIDDs. When traversing the edges from the top of a QuIDD to a terminal value, each multiplicative edge simply multiples the terminal at the end of the path by its value. Consider again the

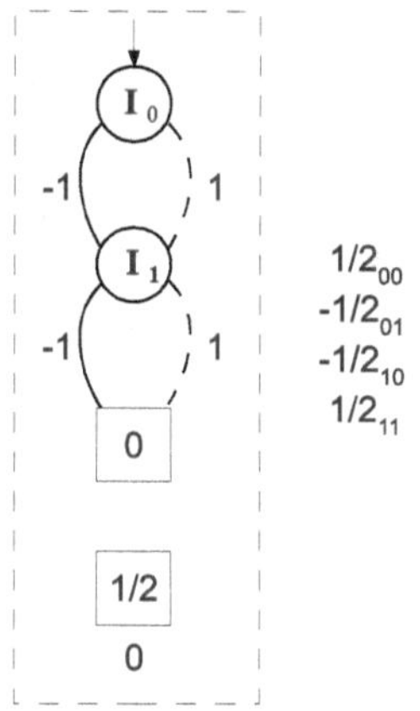

Fig. 7.6. Multiplicative edge-valued QuIDD that reduces the number of nodes of the QuIDD shown in Figure 7.1c. The values on the 1 edges imply multiplication by -1. Tracing each path produces the state vector shown on the right.

state vector QuIDD shown in Figure 7.1c. The number of nodes in this QuIDD can be reduced by using multiplicative edge values. Figure 7.6 provides one such reduction.

Generally speaking, for any unentangled (separable) state, a multiplicative edge-valued QuIDD can be constructed by simply creating one internal node for each qubit, setting each outgoing edge value to a complex number from the 2-element state vector, and connecting the next qubit's internal node to both edges. In this construction, only one terminal node exists, and its value is 1. An example of a general construction for such a QuIDD is shown in Figure 7.7. The state represented by this QuIDD is $|\psi\rangle = \begin{bmatrix} a \\ b \end{bmatrix} \otimes \begin{bmatrix} c \\ d \end{bmatrix} \otimes \begin{bmatrix} e \\ f \end{bmatrix}$.

Figures 7.3a and 7.3b demonstrate why the multiplicative edge-valued construction is so intuitive. As discussed earlier,the tensor product operation $A \otimes B$ reduces to recursively chaining the entire graph of B to the terminal nodes of A, where the terminal nodes of A simply become the edge values that connect A to B. Since redundant nodes are not maintained in QuIDDs, all of the terminal nodes of A become edges that connect to the head node of B. Figure 7.8 shows the multiplicative edge-valued version of Figure 7.3a. Notice that the only difference between the two constructions is that C_1 and C_2 become edge values that both connect to H^{n-1}.

Efficient representations of unentangled states is not a unique feature of this representation, and can be found in Vidal's slightly-entangled simulation technique [106] introduced in Section 6.4. Like Vidal's technique, applying one-qubit gates to multiplicative edge-valued QuIDDs representing separable states[1] does not increase the size of the representation. In fact, the application

[1] A general 1-qubit **Apply**-based algorithm is described later in Section 10.1.

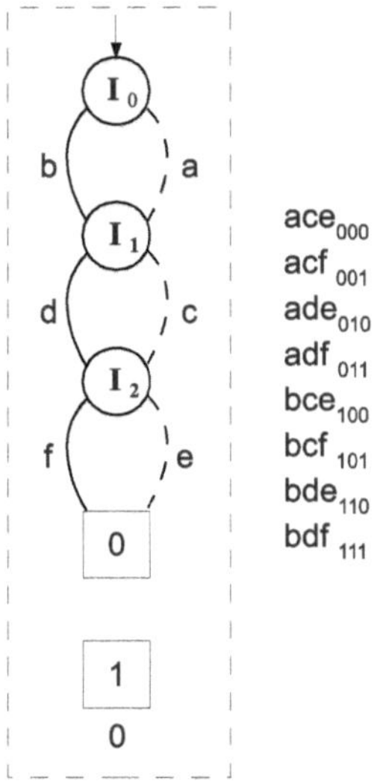

Fig. 7.7. General construction of a multiplicative edge-valued QuIDD for a 3-qubit separable state and corresponding state vector.

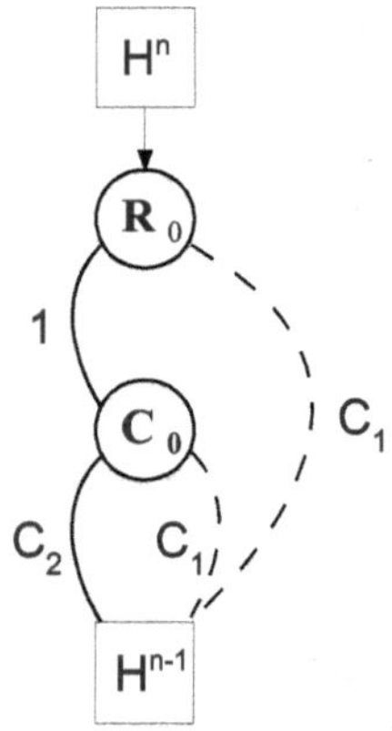

Fig. 7.8. Multiplicative edge-valued QuIDD representation of n-qubit Hadamard QuIDD from Figure 7.3a. As before, $C_1 = 1/\sqrt{2}$ and $C_2 = -1/\sqrt{2}$.

of any 1-quit gate only changes the edge values of the internal node corresponding to the affected qubit. This should be an intuitive result since one could apply a 1-qubit operator's 2×2 matrix to the 2-element state vector representing the affected qubit and then reconstruct a multiplicative edge-valued QuIDD using the above procedure. The construction always produces one internal node per qubit.

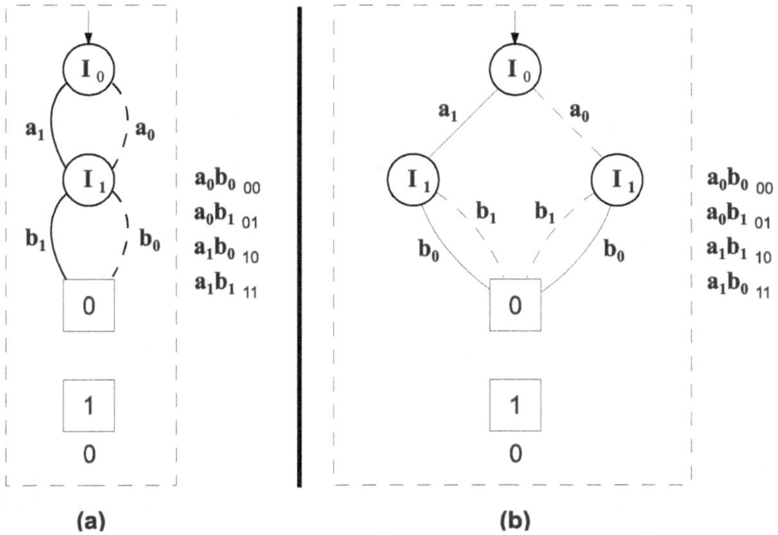

Fig. 7.9. Multiplicative edge-valued QuIDDs for (a) a separable 2-qubit state and (b) the resulting inseparable state after application of a CNOT operator.

A major problem arises however when 2-qubit gates are applied that introduce entanglement. For example, consider applying a CNOT gate to the state $|AB\rangle = \begin{bmatrix} a_0 \\ a_1 \end{bmatrix} \otimes \begin{bmatrix} b_0 \\ b_1 \end{bmatrix}$, where A is the control and B is the target. Figures 7.9a and 7.9b show the before and after states resulting from the application of the CNOT gate. Notice that a split is introduced at the internal node representing B. In general, CNOT gates alone can expand a multiplicative edge-valued QuIDD into a full binary tree. This is not surprising since the set of all 1-qubit gates and the CNOT gate represent a universal gate library for quantum computing. Since 1-qubit gates do not increase the size of a QuIDD, one would naturally expect the CNOT gate to have the potential to exponentially increase the size of a QuIDD in the worst case.

The problem becomes even more complicated for arbitrary gates that act on two or more qubits. Even restricting the allowable gates to the universal set of one-qubit and CNOT gates, using the decomposition techniques from [79] leads to the same problem due to the CNOT gate. It is not obvious how to extend the **Apply**-based matrix multiplication to handle such gates in a way that consistently assigns edge values to the resulting QuIDD, and not only maintains canonicity, but also chooses the edge values to maximize compression. Figures 7.10a and 7.10b demonstrate this problem by showing two multiplicative edge-valued QuIDDs that represent the same state using different edge values. Notice that Figure 7.10b achieves greater compression due to its choice of edge values. Key challenges with such data structures

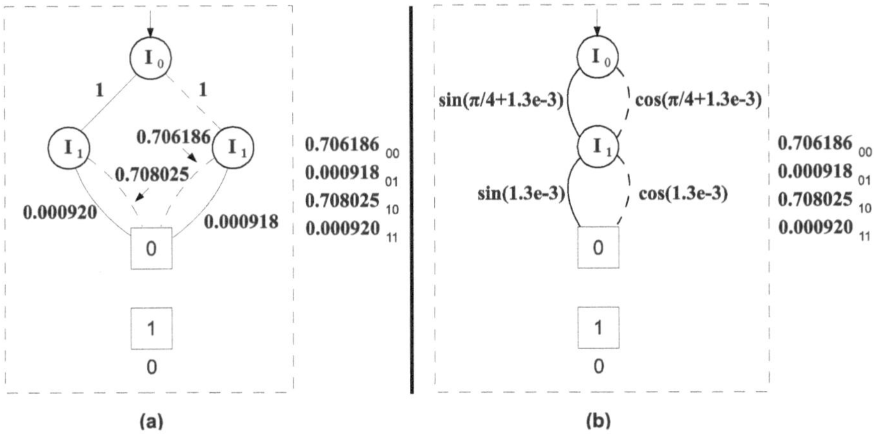

Fig. 7.10. The effect of different edge value choices on compression: (a) a bad choice and (b) a smarter choice that takes into account trigonometric identities. All explicit values are shown to five significant digits.

include ensuring canonicity (required to share equivalent portions of the DD) by careful assignment of edge values.

Work has been done on these problems however, resulting in two new DD techniques for quantum circuit representation [3, 58]. Although there is still no obvious algorithm for choosing the best edge values, these new techniques address the important problem of reconciling edge values with canonicity. This work is reviewed in the next two subsections.

Quantum Multiple-valued Decision Diagrams

These decision diagrams, referred to as QMDDs, focus on the representation of the gates in quantum circuits using multiplicative edge values and specialized **Apply**-based algorithms for matrix multiplication, the tensor product, and initial construction [58]. Two key structural extensions developed for QMDDs are multiplicative edge values as described previously and the generalization to quantum gates that can operate on *qudits* which are quantum digits with $d = 2, 3, 4, \ldots$ possible values.

To ensure canonicity, QMDDs employ the following rules [58]:

1. Each internal node for d-valued logic has d^2 outgoing edges augmented with complex-valued labels.
2. The edges of each internal node are normalized.
3. A single terminal node is used with a value fixed at 1.
4. For a given variable ordering $x_0 \prec x_1 \prec \ldots \prec x_{n-1}$, each path through the QMDD from top to bottom visits internal nodes in the reverse of the variable ordering.

5. Like most other BDD-based data structures, no internal node is redundant. This is a key difference from the multiplicative edge-valued QuIDDs described in the previous section.
6. Internal nodes are unique in that no two internal nodes contain the same outgoing edge values and destination nodes.

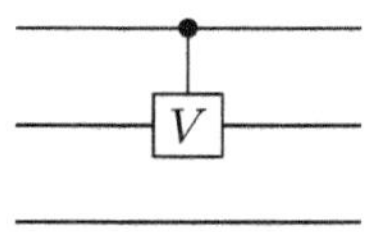

Fig. 7.11. Graphical symbol for a controlled-V gate acting on the first two qubits.

To illustrate the QMDD structure, consider the 2-qubit controlled-V gate in the circuit of Figure 7.11, and the circuit's corresponding canonical QMDD representation in Figure 7.12. The V gate has the matrix representation $V = \frac{1+i}{2}\begin{bmatrix} 1-i \\ -11 \end{bmatrix}$. Since the edge values are multiplicative, the intuition behind how the QMDD maps to the equivalent matrix representation is very similar. Compare the QMDD in Figure 7.12 to the explicit matrix representation for the circuit [58]:

$$\begin{bmatrix} 1 & 0 & 0 & 0 & 0 & 0 & 0 & 0 \\ 0 & 1 & 0 & 0 & 0 & 0 & 0 & 0 \\ 0 & 0 & 1 & 0 & 0 & 0 & 0 & 0 \\ 0 & 0 & 0 & 1 & 0 & 0 & 0 & 0 \\ 0 & 0 & 0 & 0 & \frac{1+i}{2} & 0 & \frac{1-i}{2} & 0 \\ 0 & 0 & 0 & 0 & 0 & \frac{1+2}{2} & 0 & \frac{1-i}{2} \\ 0 & 0 & 0 & 0 & \frac{1-i}{2} & 0 & \frac{1+i}{2} & 0 \\ 0 & 0 & 0 & 0 & 0 & \frac{1-i}{2} & 0 & \frac{1+i}{2} \end{bmatrix}$$

QMDDs have been shown to compress certain quantum circuits very well. For example, a QMDD represents the reversible circuit benchmark called *cycle17_3* [56] using only 236 nodes and is constructed in less than 0.1 seconds on a common type of workstation. This is an impressive result considering that this 20-qubit circuit contains 48 Toffoli gates, and naive simulation would require 48 multiplications of various $2^{20} \times 2^{20}$ matrices.

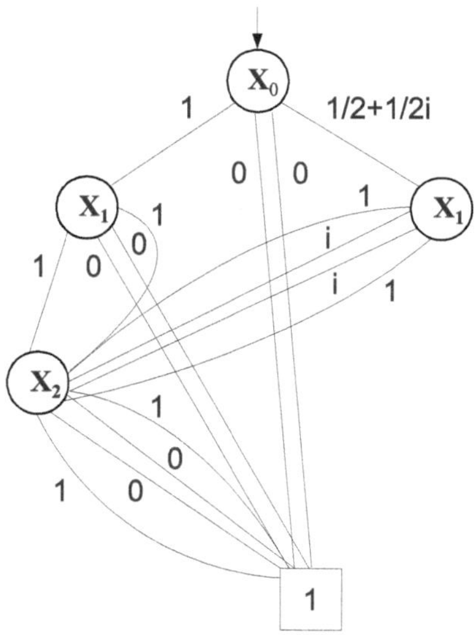

Fig. 7.12. QMDD representation of the controlled-V gate circuit from Figure 7.11.

Quantum Decision Diagrams

QDDs[2] focus exclusively on automatic synthesis using controlled and uncontrolled $R_x(\theta)$ gates [3]. (See Chapter 4 for a description of rotation gates such as R_x). They are similar to traditional ROBDDs with the following enhancements:

1. By restricting the allowable operators to $R_x(\theta)$ gates, QDDs introduce a novel edge value which represents the rotation angle θ.
2. Internal variables represent qubits.
3. Traversing the 1 edge with value θ from an internal node with variable v_i to another internal node with variable v_{i+1} represents applying a controlled $R_x(\theta)$ operator to v_{i+1} with v_i as the control.
4. Traversing the 0 edge from an internal node with variable v_i to another internal node with variable v_{i+1} applies no operator to v_{i+1}. As a result, the 0 edges contain no edge values.
5. Unlike QMDDs, but like multiplicative edge-valued QuIDDs, QDDs may contain nodes whose outgoing edge values all connect to the same node.

A sample QDD is shown in Figure 7.13b. In this case, the QDD represents the controlled $R_x(\pi)$ gate with three controls as shown in Figure 7.13a. Notice that the edges connecting control qubit nodes have no edge labels. This is equivalent to applying $R_x(0)$ which is merely the identity operator.

[2] These QDDs should not be confused with Greve's earlier QDD implementation for simulating Shor's algorithm [37], which is a completely different data structure.

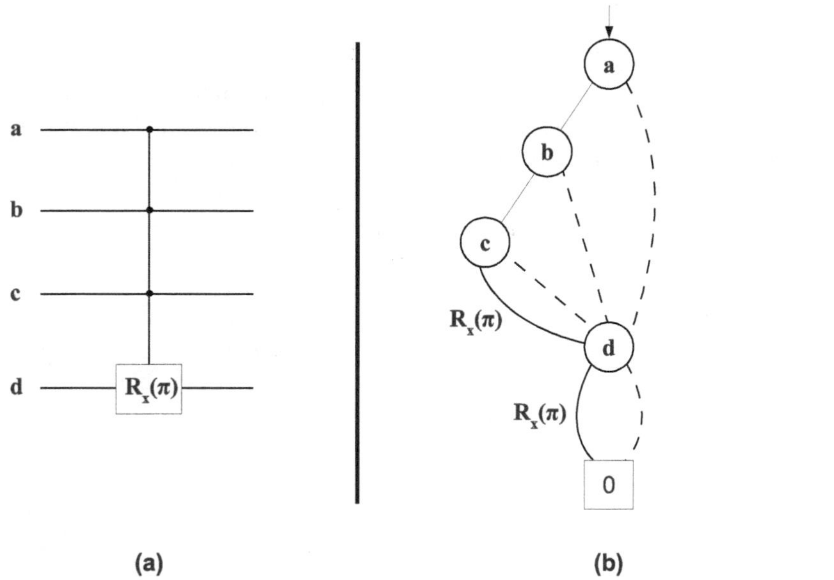

Fig. 7.13. (a) Circuit depiction of a controlled-$R_x(\pi)$ gate with three controls, and (b) the corresponding QDD representation of the circuit.

Like QMDDs, QDDs are canonical [3], making them well-suited for quantum circuit synthesis where checking the equivalence of potentially different circuits is a common procedure. Interestingly, QDDs have been shown to reproduce the optimal constructions of some elementary quantum circuit operators [3, 7]. In such experiments, the functionality of the operator to be synthesized is provided explicitly as a full binary tree version of a QDD. A special algorithm is then applied to reduce the QDD with exponentially many nodes in the number of qubits to a potentially asymptotically smaller QDD. In other words, a typical way to construct a QDD is first to represent the operator's function as a QDD that violates some of the QDD reduction rules, and subsequently legalize the QDD by applying those rules recursively.

Creating QDDs explicitly using trees with exponentially many nodes prohibits the synthesis of circuits with non-trivial numbers of qubits. As a result, QDDs offer an **Apply**-based algorithm which allows new $R_x(\theta)$ gates to be incorporated into a QDD when a circuit is read in gate-by-gate. Like QuIDDs, various elementary gates may be created with the explicit technique since the size of the gates is small, and circuits may then be subsequently constructed by combining these smaller QDDs using the **Apply**-based algorithm.

7.5 Summary

We have examined a DD-based technique for simulating quantum circuits using a data structure called a QuIDD. QuIDDs enable practical, generic and reasonably efficient simulation of quantum computation. Their key advantages are fast execution and low memory usage. In our experiments, QuIDDPro achieves exponential memory savings compared to other known techniques. We have also reviewed some newer ways to apply DDs to quantum circuit simulation and synthesis. These approaches are distinguished by various different types of edge values and reduction rules.

In a more general sense, this work explores some of the limitations of quantum computing. Classical computers have the advantage that they are not subject to quantum measurement and errors. Thus, when competing with quantum computers, classical computers can simply run ideal error-free quantum algorithms, allowing techniques such as QuIDDs, QMDDs and QDDs to exploit the symmetries found in ideal quantum computation. On the other hand, quantum computation still has certain operators which cannot be represented using only polynomial resources on a classical computer, even with efficient DD-based techniques, e.g., the quantum Fourier transform (QFT) and its inverse, which are used in Shor's number factoring algorithm [82]. Figure 7.14 shows the growth in number of nodes of the $N \times N$ inverse QFT as a QuIDD. Since $N = 2^n$ where n is the number of qubits, this QuIDD exhibits exponential growth with a linear increase in qubits. Therefore, the inverse QFT will cause QuIDDPro to have exponential runtime and memory requirements when simulating Shor's algorithm.

Another challenging aspect of quantum simulation is the impact of errors due to defects in circuit components, and environmental effects such as decoherence. Error simulation appears to be essential for modeling actual quantum computational devices. This may, however, prove to be difficult since errors can alter the symmetries exploited by QuIDDs and other DD techniques. One way to handle errors is to extend QuIDDs to encompass the density matrix representation; this extension is described in the next chapter.

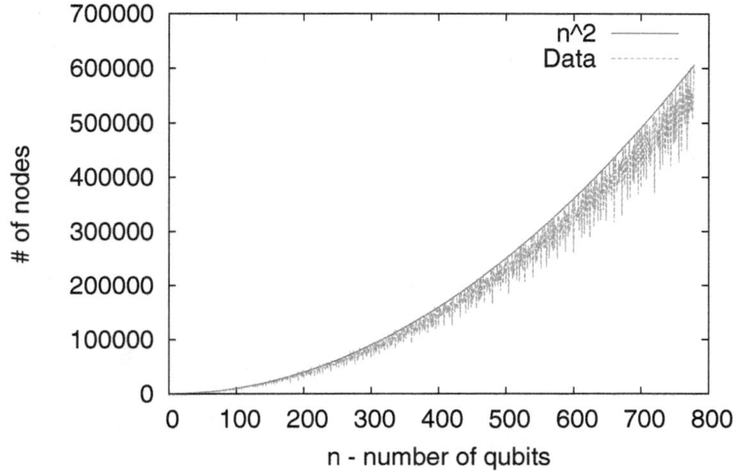

Fig. 7.14. Growth of inverse quantum Fourier transform matrix in QuIDD form.

8

Density-Matrix Simulation with QuIDDs

This chapter extends QuIDD-based quantum circuit simulation to handle density matrices, which are crucial in capturing interactions between quantum states and their environment. Besides the usual operations required to simulate with the state-vector model, including matrix multiplication and the tensor product, simulation with density matrices requires the outer product and the partial trace. We therefore present algorithms to implement the outer product and the partial trace with QuIDDs. The outer product is used in the initialization of density matrices. The partial trace is invaluable in error modeling as it facilitates descriptions of single qubit states that are affected by noise and other phenomena. We also describe a set of quantum circuit benchmarks that incorporate errors, error correction, reversible logic, quantum communication, and quantum search. To evaluate the improvements offered by the QuIDD-based approach empirically, we use these benchmarks to compare QuIDDPro with an array-based density matrix simulator called QCSim that makes extensive use of qubit-wise multiplication.

8.1 QuIDD Properties and Density Matrices

Although the density matrix representation is invaluable for simulating environmental noise in quantum circuits, like the state vector representation, it is plagued by runtime and memory complexity that can grow exponentially with the number of qubits. As discussed in Section 3.2, a straightforward linear-algebraic simulation using density matrices requires $O(2^{2n})$ time and memory resources. Since QuIDDs have proven useful in reducing this complexity in the state vector paradigm, it is only natural to extend QuIDDs to the density matrix model in an attempt to reduce the simulation complexity of this important model in practical cases. Before proceeding to the new extensions, it is instructive to first review what is already in place that can be re-used with density matrces.

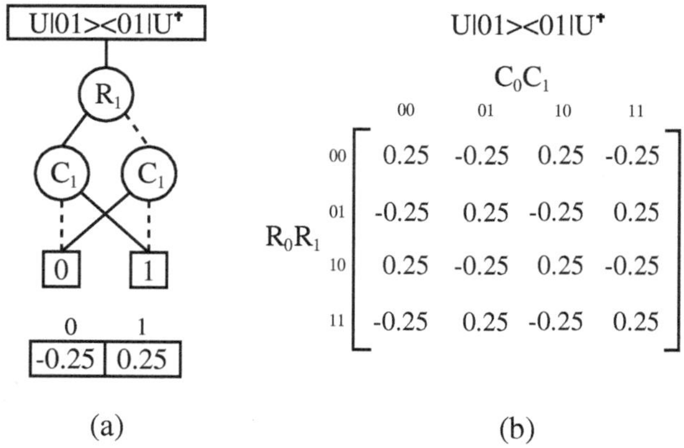

Fig. 8.1. (a) QuIDD for the density matrix resulting from $U|01\rangle\langle 01|U^{\dagger}$, where $U = H \otimes H$, and (b) its explicit matrix form.

Figure 8.1a shows the QuIDD that results from applying $U = H \otimes H$ to an outer product thus: $U|01\rangle\langle 01|U^{\dagger}$. The R_i nodes of the QuIDD encode the binary indices of the rows in the corresponding explicit matrix (Figure 8.1b). Similarly, the C_i nodes encode the binary indices of the columns. Solid lines leaving a node denote the positive cofactor of the index bit variable (a value of 1), while dashed lines denote the negative cofactor (a value of 0). Terminal nodes correspond to the value of the element in the explicit matrix whose binary row/column indices are encoded by the path that was traversed through the QuIDD.

Notice that the first and third rows of the explicit matrix in Figure 8.1b are the same, as are the first and third columns. This redundancy is captured by the QuIDD in Figure 8.1a because the QuIDD does not contain any R_0 or C_0 nodes. In other words, the values and their locations in the explicit matrix can be completely determined without the superfluous knowledge of the first row and column index bits.

Measurement, matrix multiplication, addition, scalar products, the tensor product, and other operations involving QuIDDs are variations of the basic **Apply** algorithm (Chapter 7). Vectors and matrices with large blocks of repeated values can be manipulated in QuIDD form quite efficiently with these operations. Section 7.2 identifies a class of vectors and matrices that is simulated efficiently with QuIDDs. Since QuIDDs already are able to represent matrices and multiply them, extending QuIDDs to encompass the density matrix representation requires algorithms for the outer product and the partial trace.

```
Outer_Product(Q,num_qubits) {
  Q_cctrans = Swap_Row_Col_Vars(Q)
  Q_cctrans = Complex_Conj(Q_cctrans)
  R = Matrix_Multiply(Q,Q_cctrans)
  R = Scalar_Div(Q_cctrans,2^num_qubits)
  return R
}
```

(a)

```
Complex_Conj(Q) {
  if (Is_Constant(Q))
    return New_Terminal(
      real(Q), -1 * imag(Q))
  if (Table_Lookup(R,Q)
    return R
  v = Top_Var(Q)
  T = Complex_Conj(Q_v)
  E = Complex_Conj(Q_v')
  R = ITE(v,T,E)
  Table_Insert(R,Q)
  return R
}
```

(b)

Fig. 8.2. (a) The QuIDD outer product algorithm and (b) its complex conjugation helper function `Complex_Conj`. `Scalar_Div` is the same as `Complex_Conj` except that it returns the value of the terminal divided by a scalar.

8.2 QuIDD-based Outer Product

The outer product involves matrix multiplication between a column vector and its complex-conjugate transpose. Since a column vector QuIDD only depends on row variables, the transpose can be accomplished by swapping the row variables with column variables. The complex conjugate can then be performed with a DFS traversal that replaces terminal node values with their complex conjugates. The original column vector QuIDD is then multiplied by its complex-conjugate transpose using the matrix multiply operation previously defined for QuIDDs (Section 7.1).

Pseudocode for the resulting algorithm is given in Figure 8.2. Notice that before the result is returned, it is divided by 2^{num_qubits}, where num_qubits is the number of qubits represented by the QuIDD vector. This is done because a QuIDD that only depends on n row variables can be viewed as either a $2^n \times 1$ column vector or a $2^n \times 2^n$ matrix in which all columns are the same. Since matrix multiplication is performed according to the latter case [99, 101, 6], the result of the outer product contains values that are multiplied by an extra factor of 2^n, and therefore must be normalized.

Although QuIDDs enable efficient simulation for a class of matrices and vectors in the state-vector paradigm, it must be shown that the density-matrix version of this class can also be simulated efficiently. Since state vectors are converted to density matrices via the outer product, this requires proving that the outer product of a QuIDD vector in this class with its complex-conjugate transpose results in a QuIDD density matrix of size polynomial in the number of qubits.

Theorem 8.1. *Given an n-qubit QuIDD state vector created from tensor products of QuIDDs with $O(1)$ nodes whose terminal values are in a persistent set, the outer product of this QuIDD with its complex-conjugate transpose produces a QuIDD matrix with polynomially many nodes in n.*

Proof. Since the given QuIDD state vector's terminal values are in a persistent set, the number of nodes in the QuIDD is $O(n)$ (Theorem 7.7). Consider the pseudocode for the QuIDD outer product shown in Figure 8.2a. The first operation is to create a transposed copy of the QuIDD state vector. Transposition only requires remapping the internal variable nodes to represent column variables instead of row variables. This can be done in one pass over all the nodes in the QuIDD state vector (Section 7.1). Since the number of nodes is $O(n)$, this operation has $O(n)$ runtime complexity and creates a transposed copy with $O(n)$ nodes. The next operation is to form the complex conjugate of the transposed QuIDD copy. As evidenced by the pseudocode for complex conjugation of QuIDDs in Figure 8.2b, this involves a single recursive pass over all nodes. All internal nodes are returned unchanged with the $O(1)$ ADD ITE operation [6], whereas the complex conjugates of the terminals are returned when they are reached. Since the number of nodes in the transposed QuIDD copy is $O(n)$, the runtime complexity of this operation is $O(n)$ and results in a new QuIDD with $O(n)$ nodes.

Next, QuIDD matrix multiplication is performed on the QuIDD state vector and its complex-conjugate transpose to produce the QuIDD density matrix. It has been proven that QuIDD matrix multiplication of some QuIDD A with $|A|$ nodes and another QuIDD B with $|B|$ nodes has runtime complexity $O((|A||B|)^2)$ and results in a QuIDD with $O((|A||B|)^2)$ nodes (Section 7.1). Since the QuIDD state vector and its complex-conjugate transpose have $O(n)$ nodes, the matrix multiplication step has runtime complexity $O(n^4)$. The final normalization step of the outer product is a scalar division of the terminal values. Like QuIDD complex conjugation, this operation is implemented by a single recursive pass over the QuIDD, but when the terminals are reached, the scalar division result is returned. Since the QuIDD density matrix has $O(n^4)$ nodes, this operation has runtime complexity $O(n^4)$. Based on the complexity of all steps in the QuIDD outer product algorithm, the overall runtime complexity of the QuIDD outer product is $O(n^4)$ and results in a QuIDD density matrix with $O(n^4)$ nodes. □

8.3 QuIDD-based Partial Trace

To motivate the QuIDD-based partial trace algorithm, we note how the partial trace can be performed with explicit matrices. The trace of a matrix A is the sum of A's diagonal elements. To perform the partial trace over a particular qubit in an n-qubit density matrix, the trace operation can be applied iteratively to sub-matrices of the density matrix. Each sub-matrix is composed of four elements with row indices $r0s$ and $r1s$, and column indices $c0d$ and $c1d$, where r, s, c, and d are bit-sequences that index the n-qubit density matrix.

Tracing over the sub-matrices has the effect of reducing the dimensionality of the density matrix by one qubit. A well-known ADD operation to reduce the dimensionality of a matrix is the **Abstract** operation [6]. Given an arbitrary ADD f, abstraction of variable x_i eliminates all internal nodes of f that represent x_i by combining the positive and negative cofactors of f with respect to x_i using some binary operation. In other words, $\textbf{Abstract}(f, x_i, op) = f_{x_i}\ op\ f_{x_i'}$.

For QuIDDs, there is a one-to-one correspondence between a qubit on wire i (wires are labeled top-down starting at 0) and variables R_i and C_i. So at first glance, one might suspect that the partial trace of qubit i in f can be achieved by performing $\textbf{Abstract}(f, R_i, +)$ followed by $\textbf{Abstract}(f, C_i, +)$. However, this will sum the rows determined by qubit i independently of the columns. The desired behavior is the diagonal addition of sub-matrices while accounting for both the row and column variables due to i *simultaneously*.

The pseudocode to perform the partial trace correctly is given in Figure 8.3. Comparing this with the pseudocode for the **Abstract** algorithm [6], we see that when R_i corresponding to qubit i is reached, we take the positive and negative cofactors *twice* before making the recursive call. Since the interleaved variable ordering of QuIDDs guarantees that C_i immediately follows R_i [99, 101], taking the positive and negative cofactors twice simultaneously abstracts both the row and column variables for qubit i, thus achieving the desired goal of summing diagonals. In other words, for a QuIDD f, the partial trace over qubit i is $\textbf{Ptrace}(f, i) = f_{R_i C_i} + f_{R_i' C_i'}$. Note that the pseudocode has checks for the special case when no internal nodes in the QuIDD represent C_i. Not shown in the pseudocode is book-keeping that shifts up the variables in the resulting QuIDD to fill the hole in the ordering left by the row and column variables that were traced over.

As in the case of the outer product, the QuIDD partial trace algorithm has efficient runtime and memory complexity in the size of the QuIDD being traced over, as we now show.

Theorem 8.2. *Given an n-qubit density matrix QuIDD A with $|A|$ nodes, any qubit in the matrix can be traced over with runtime complexity $O(|A|)$ and results in a density matrix QuIDD with $O(|A|)$ nodes.*

Proof. Consider the pseudocode for the QuIDD partial trace algorithm in Figure 8.3. The algorithm performs a recursive traversal of the nodes in the QuIDD density matrix and takes certain actions when special cases are encountered. If a node is encountered that corresponds to a qubit preceded by the traced-over qubit in the variable ordering,[1] then recursion stops and the

[1] Recall the one-to-one correspondence between qubit i and variables R_i and C_i.

```
Ptrace(Q, qubit_index) {
  if(Is_Constant(Q))
    return Q
  top_q = Top_Var
  if (qubit_index < Index(top_q)) {
    R = Apply(Q, Q, +)
    return R
  }
  if (Table_Lookup(R, Q, qubit_index)
    return R
  T = Q_top_q
  E = Q_top_q'
  if (qubit_index == Index(top_q)) {
    if (Is_Constant(T) or Index(T) > Index(Q) + 1)
      r1 = T
    else {
      top_T = Top_Var(T)
      r1 = T_topT
    }
    if (Is_Constant(E) or Index(E) > Index(Q) + 1)
      r2 = E
    else {
      top_E = Top_Var(E)
      r2 = E_top_E'
    }
    R = Apply(r1, r2, +)
    Table_Insert(R, Q, qubit_index)
    return R
  }
  else {  /* (qubit_index > Index(top_q)) */
    r1 = Ptrace(T, qubit_index)
    r2 = Ptrace(E, qubit_index)
    R = ITE(top_q, r1, r2)
    Table_Insert(R, Q, qubit_index)
    return R
  }
}
```

Fig. 8.3. The QuIDD partial trace algorithm. The index of the qubit being traced over is `qubit_index`.

sub-graph processed (added to itself) with the ADD **Apply** algorithm [20]. This operation has runtime complexity $O(|A|)$ and results in a new sub-graph with $O(|A|)$ nodes. Next, if the partial trace of the current sub-graph has already been computed, then recursion stops and the pre-computed result is simply looked up in the computed table cache and returned. This operation has runtime complexity $O(1)$ and returns a sub-graph with $O(|A|)$ nodes [20]. If there is no entry in the computed table cache, the algorithm checks whether the current node's variable corresponds to the qubit to be traced over. If so, **Apply** is used to add the node's children or children's children, which again has $O(|A|)$ runtime and memory complexity. If the current node does not correspond to the qubit being traced over, then the partial trace algorithm is called recursively on the node's children. Since all the other special cases stop recursion and involve an **Apply** operation, then the overall runtime complexity of the partial trace algorithm is $O(|A|)$ and results in a new QuIDD with $O(|A|)$ nodes. □

8.4 Empirical Validation

We consider a number of quantum circuit benchmarks which cover errors, error correction, reversible logic, communication, and quantum search. We devised some of these benchmarks, while others are drawn from NIST [14] and a web site devoted to reversible circuits [56]. For every benchmark, the simulation performance of QuIDDPro is compared with NIST's QCSim quantum circuit simulator, which utilizes an explicit array-based computational engine. The results indicate that QuIDDPro far outperforms QCSim. All experiments are performed on a 1.2GHz AMD Athlon workstation with 1GB of RAM running Linux.

Reversible Circuits

First we examine the performance of QuIDDPro simulating a set of reversible circuits, which are quantum circuits that implement classical operations [61]. Specifically, if the input qubits of a quantum circuit are all in the computational basis, i.e., they have only $|0\rangle$ or $|1\rangle$ values, there is no quantum noise, and all the gates are "k-CNOT" gates with $k = 0$ for NOT, $k = 1$ for CNOT, etc. [79], then the output qubits and all intermediate states will also be in the computational basis. Such a circuit results in a classical logic operation that is reversible in the sense that the input vectors can always be derived from the output vectors and the circuit function. Reversibility comes from the fact that all quantum operators must be unitary and so have inverses [61].

The first benchmark is the reversible 4-bit ripple-carry adder depicted in Figure 8.4. Since QuIDD size is sensitive to the arrangement of different values of the matrix elements, we simulate the adder with various input values ("rc_adder1" through "rc_adder4"). This is also done for the remaining benchmarks. The two other reversible benchmarks we simulate contain fewer qubits but more gates than the ripple-carry adder. One of these is a 12-qubit circuit that outputs a $|1\rangle$ on the last qubit if and only if the number of $|1\rangle$'s in the input qubits is 3, 4, 5 or 6 ("9sym1" through "9sym5") [56]. The other benchmark is a 15-qubit reversible circuit that generates the classical Hamming code of the input qubits ("ham15_1" through "ham15_3") [56].

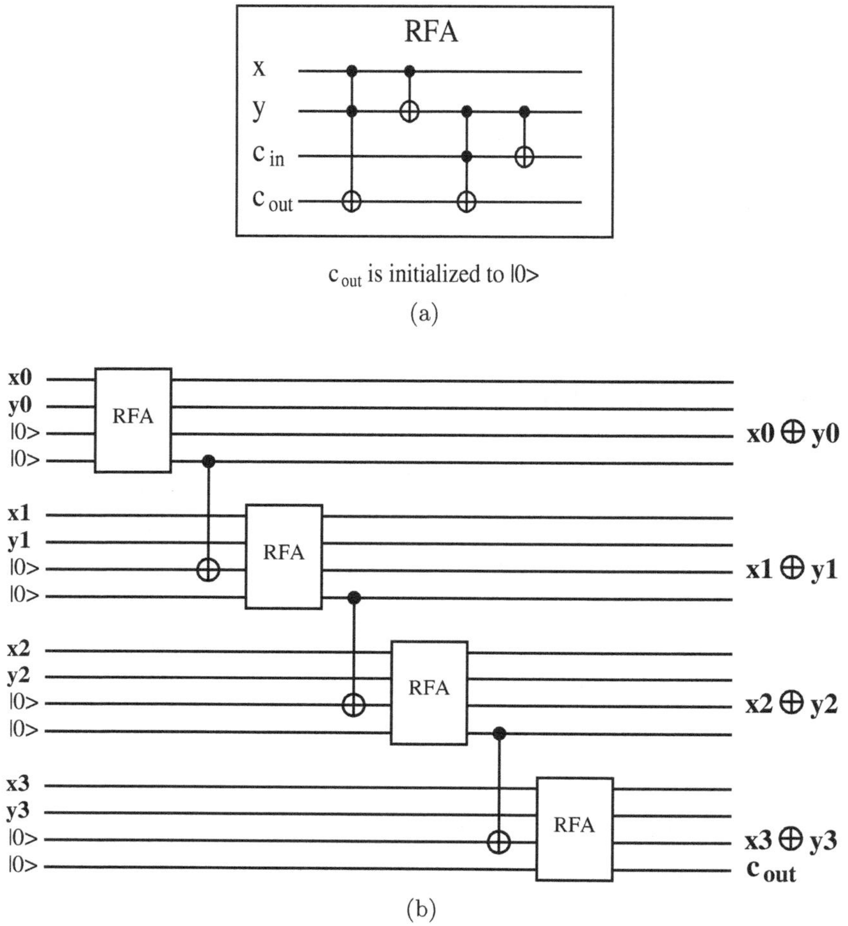

Fig. 8.4. (a) An implementation of a reversible full-adder (RFA), and (b) a reversible 4-bit ripple-carry adder that has the RFA as a module. The adder computes four binary sum bits $x_i \oplus y_i$ and an output carry bit c_{out}.

Performance results for all of these benchmarks are shown in Table 8.1. As can be seen, QuIDDPro significantly outperforms QCSim in every case. In fact, for circuits of 14 or more qubits, QCSim requires more than 2GB of memory. Since QCSim uses an explicit array-based simulation engine, it is insensitive to the arrangement and values of elements in matrices. Therefore, one can expect QCSim to use more than 2GB of memory for *any* benchmark with 14 or more qubits, regardless of circuit function or input values. Another interesting result is that even though QuIDDPro is, in general, sensitive to the arrangement and values of matrix elements, the data indicate that QuIDDPro is insensitive to varied inputs on the same circuit for error-free reversible

Benchmark	No. of qubits	No. of gates	QuIDDPro Runtime (s)	QuIDDPro Peak memory (MB)	QCSim Runtime (s)	QCSim Peak memory (MB)
rc_adder1	16	24	0.44	0.0625	—	$> 2GB$
rc_adder2	16	24	0.44	0.0625	—	$> 2GB$
rc_adder3	16	24	0.44	0.0625	—	$> 2GB$
rc_adder4	16	24	0.44	0.0625	—	$> 2GB$
9sym1	12	29	0.2	0.0586	8.01	128.1
9sym2	12	29	0.2	0.0586	8.02	128.1
9sym3	12	29	0.2	0.0586	8.04	128.1
9sym4	12	29	0.2	0.0586	8	128.1
9sym5	12	29	0.2	0.0586	7.95	128.1
ham15_1	15	148	1.99	0.121	—	$> 2GB$
ham15_2	15	148	2.01	0.121	—	$> 2GB$
ham15_3	15	148	1.99	0.121	—	$> 2GB$

Table 8.1. Performance results for QuIDDPro and QCSim on the reversible circuit benchmarks. "$> 2GB$" indicates that memory usage of 2GB was exceeded.

benchmarks. However, QuIDDPro still compresses the tremendous amount of redundancy present in these benchmarks.

Error Correction and Communication

Now we analyze the performance of QuIDDPro on simulations that incorporate errors and error correction. We consider some simple benchmarks that encode single qubits into Steane's 7-qubit error-correcting code [87], and some more complex benchmarks that use the Steane code to correct a combination of bit-flip and phase-flip errors in a half-adder and Grover's quantum search algorithm [38]. Secure quantum communication is also considered here because eavesdropping disrupts a quantum channel and so can be treated as a type of error.

The first two benchmarks "steaneX" and "steaneZ" encode a single logical qubit as seven physical qubits with the Steane code, and simulate the effect of a probabilistic bit-flip and phase-flip error, respectively [14]. "steaneZ" contains 13 qubits which are initialized to the state $0.866025|0000000000000\rangle + 0.5|0000001000000\rangle$. A combination of gates apply a probabilistic phase-flip to one of the qubits and calculate the error syndrome and error rate. "steaneX" is a 12-qubit version of the same circuit that simulates a probabilistic bit-flip error.

A more complex benchmark simulated is a reversible half-adder with three logical qubits that are encoded into 21 physical qubits by the Steane code. Additionally, three ancillary qubits are used to track the error rate, giving a total circuit size of 24 qubits. The benchmarks "hadder1_bf1" through "hadder3_bf3" simulate the half-adder with different numbers of bit-flip errors on

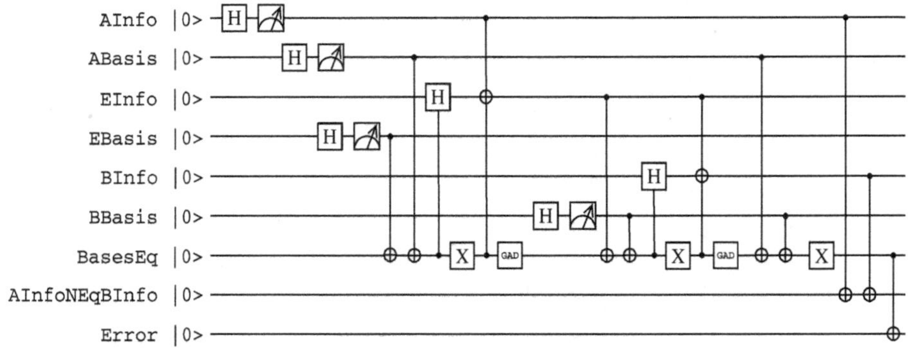

Fig. 8.5. Quantum circuit for the "bb84Eve" benchmark.

various physical qubits in the encoding of one of the logical qubit inputs. Similarly, "hadder1_pf1" through "hadder3_pf3" simulate the half-adder with various phase-flip errors.

Another large benchmark used is an instance of Grover's quantum search algorithm [38]. Whereas the simulations of this algorithm described in the previous subsection used state vectors, this experiment uses the density-matrix representation. The oracle in this case searches for one element in a database of four items. It has two logical data qubits and one ancillary oracle qubit, all of which are encoded with the Steane code. Like the half-adder circuit, this results in a total circuit size of 24 qubits. "grover_s1" simulates the circuit with the encoded qubits in the absence of errors. "grover_s_bf1" and "grover_s_pf1" introduce and correct a bit-flip and phase-flip error, respectively, on one of the physical qubits in the encoding of the oracle qubit.

An important application of quantum circuits is secure communication using quantum cryptography. The basic concept here is to use entanglement to distribute a shared key. Eavesdropping constitutes a measurement of the quantum state representing the key, and so disrupts that state. Such disruption can be detected by the communicating parties. Since actual implementations of quantum key distribution have already been demonstrated [27], efficient simulation of these protocols may be useful in exploring possible improvements. Therefore, we present two benchmarks which implement BB84, one of the earliest quantum key distribution protocols [9]. "bb84Eve" accounts for the case in which an eavesdropper is present (see Figure 8.5) and contains 9 qubits, whereas "bb84NoEve" accounts for the case in which no eavesdropper is present and contains 7 qubits. In both circuits, all qubits are traced over at the end except for two qubits reserved to track whether or not the legitimate communicating parties successfully shared a key (BasesEq) and the error due to eavesdropping (Error).

Performance results for the benchmarks are shown in Table 8.2. Again, QuIDDPro significantly outperforms QCSim on all benchmarks except for "bb84Eve" and "bb84NoEve," where the performance of QuIDDPro and QC-

Benchmark	No. of qubits	No. of gates	QuIDDPro Runtime (s)	QuIDDPro Peak memory (MB)	QCSim Runtime (s)	QCSim Peak memory (MB)
steaneZ	13	143	0.6	0.672	287	512
steaneX	12	120	0.27	0.68	53.2	128
hadder_bf1	24	49	18.3	1.48	—	$> 2GB$
hadder_bf2	24	49	18.7	1.48	—	$> 2GB$
hadder_bf3	24	49	18.7	1.48	—	$> 2GB$
hadder_pf1	24	51	21.2	1.50	—	$> 2GB$
hadder_pf2	24	51	21.2	1.50	—	$> 2GB$
hadder_pf3	24	51	20.7	1.50	—	$> 2GB$
grover_s1	24	50	2301	94.2	—	$> 2GB$
grover_s_bf1	24	71	2208	94.3	—	$> 2GB$
grover_s_pf1	24	73	2258	94.2	—	$> 2GB$
bb84Eve	9	26	0.02	0.129	0.19	2.0
bb84NoEve	7	14	<0.01	0.0313	<0.01	0.152

Table 8.2. Performance results for QCSim and QuIDDPro on benchmarks incorporating errors. "$> 2GB$" indicates that a memory usage of 2GB was exceeded.

Sim is about the same. The reason is that these benchmarks contain fewer qubits than all of the others. Since each additional qubit doubles the size of an explicit density matrix, QCSim has difficulty simulating the larger Steane-encoded benchmarks.

Scalability and Quantum Search

To analyze scalability with the number of input qubits, we turn to quantum circuits that allow a variable number of inputs. In particular, we reconsider Grover's quantum search algorithm. For these experiments, the qubits are not encoded with the Steane code, and errors are not introduced. The oracle performs the function described earlier, except that the number of data qubits now ranges from 5 to 20.

Performance results for the Grover algorithm benchmarks are shown in Table 8.3. Again, QuIDDPro has significantly better performance. These results highlight the fact that QCSim's explicit representation of the density matrix becomes an asymptotic bottleneck as n increases, while QuIDDPro's compression of the density matrix and operators scales very well.

8.5 Summary

We have described a new graph-based simulation technique that enables efficient density-matrix simulation of quantum circuits. We implemented this technique in the QuIDDPro simulator, which uses the QuIDD data structure to compress redundancy in the gate operators and the density matrix. As a result, the time and memory complexity of QuIDDPro varies with the structure of the circuit. However, we demonstrated that QuIDDPro exhibits superior performance on a set of benchmarks which incorporate qubit errors, mixed states, error correction, quantum communication, and quantum search. These results indicate that there is a great deal of structure in *practical* quantum circuits that graph-based algorithms like those implemented in QuIDDPro can exploit.

No. of qubits	No. of gates	QuIDDPro		QCSim	
		Runtime (s)	Peak memory (MB)	Runtime (s)	Peak memory (MB)
5	32	0.05	0.0234	0.01	0.00781
6	50	0.07	0.0391	0.01	0.0352
7	84	0.11	0.043	0.08	0.152
8	126	0.16	0.0586	0.54	0.625
9	208	0.27	0.0742	3.64	2.50
10	324	0.42	0.0742	23.2	10.0
11	520	0.66	0.0898	151	40.0
12	792	1.03	0.105	933	160
13	1224	1.52	0.141	5900	640
14	1872	2.41	0.125	—	$> 2GB$
15	2828	3.62	0.129	—	$> 2GB$
16	4290	5.55	0.145	—	$> 2GB$
17	6464	8.29	0.152	—	$> 2GB$
18	9690	12.7	0.246	—	$> 2GB$
19	14508	18.8	0.199	—	$> 2GB$
20	21622	28.9	0.203	—	$> 2GB$

Table 8.3. Performance results for QCSim and QuIDDPro on the Grover's quantum search benchmark. "$> 2GB$"" indicates that a memory usage of 2GB was exceeded.

9

Checking Equivalence of States and Circuits

Equivalence checking is a basic task in the synthesis and verification of classical digital circuits. A hardware designer needs to know whether a circuit's implementation is functionally equivalent to its specification. In addition, the equivalence of different versions of the same (sub-)circuit must be checked throughout the complex computer-aided design process. Traditional combinational equivalence checking is solved in practice with high-performance solvers for Boolean Satisfiability, and its negative version (non-equivalence) is NP-complete.

Equivalence checking is likely to be just as important in quantum CAD, and the non-equivalence of quantum circuits is QMA-complete.[1] However, the equivalence of quantum states and operators can be subtle. Unlike their classical counterparts, qubits and quantum gates can differ by global and relative phase, and yet be equivalent upon measurement. Building upon the algorithmic blocks developed in Chapter 7, we present QuIDD algorithms to check quantum states and operators for equivalence. As we will see, the variety of algorithms available to solve this problem is surprising.

9.1 Quantum Equivalence Checking

As observed above, quantum circuits require the classical concept of equivalence to be extended to account for global and relative phase. This broader notion of equivalence creates several new opportunities in quantum circuit design, where minimizing the number of gate operations to achieve a given function is a fundamental goal. For example, the Toffoli gate can be implemented with fewer controlled-NOT (CNOT) and 1-qubit gates, if equivalence is interpreted as "equivalence up to relative phase" or, more briefly, "relative-phase equivalence" [7, 84], as shown in Figure 9.1. Normally the Toffoli gate

---

[1] QMA is the quantum version of the complexity class NP.

requires an equivalent circuit of six CNOT and eight 1-qubit gates to implement it [78]. Any relative-phase differences present in an equivalent circuit can be canceled out so long as every pair of these gates in the circuit is strategically placed [84]. Since circuit minimization is being pursued for a number of key quantum arithmetic circuits with many Toffoli gates, such as the modular exponentiation occurring in Shor's algorithm, [92, 28, 29, 77, 76], this type of phase equivalence could reduce the number of gates even further.

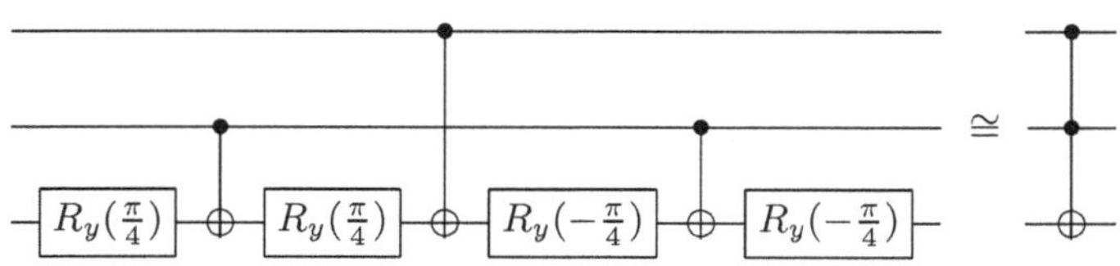

Fig. 9.1. Margolus' circuit, which is relative-phase equivalent to the Toffoli gate.

Recall that two states $|\psi\rangle$ and $|\varphi\rangle$ are equal up to global phase if $|\varphi\rangle = e^{i\theta}|\psi\rangle$, where $\theta \in \mathbb{R}$. The global phase factor $e^{i\theta}$ will not be observed on measurement of either state [61]. In contrast, two states are equal up to relative phase if

$$|\varphi\rangle = \begin{bmatrix} e^{i\theta_0} & & & \\ & e^{i\theta_1} & & \\ & & \ddots & \\ & & & e^{i\theta_{N-1}} \end{bmatrix} |\psi\rangle . \tag{9.1}$$

The probability amplitudes of the state $U|\psi\rangle$ will, in general, differ by more than relative phase from those of $U|\varphi\rangle$, but the measurement outcomes may be the same. Global-phase equivalence is a special case of relative-phase equivalence in which all $e^{i\theta_j}$ are equal. Furthermore, identical states may be considered a special case of global-phase equivalence in which the phase factor is 1. Thus, the equivalence-checking problem can be viewed as an equivalence hierarchy in which exact equivalence implies global-phase equivalence, which implies relative-phase equivalence, which in turn implies measurement-outcome equivalence. The equivalence-checking problem is also extensible to quantum operators with applications to quantum-circuit synthesis and verification, which involve the computer-aided generation of quantum circuits that have correct functionality and a minimum or near-minimum number of gates.

Equivalence relations may be expressed directly in terms of matrices and matrix operations, as in Equation 9.1. Here, however, we present QuIDD algorithms to accomplish the task more efficiently. Our algorithms solve the equivalence-checking problem asymptotically faster in some cases. Empirical results confirm the algorithms' effectiveness and show that the improvements are more significant for operators than for states. Interestingly, solving the equivalence problem for the benchmark circuits considered requires significantly less time than creating their QuIDD representations, which indicates that such problems can be efficiently solved in practice using quantum-circuit CAD tools.

9.2 Global-Phase Equivalence

This section describes algorithms that check equivalence up to global phase of two quantum states or operators. The first two algorithms are well-known linear-algebraic operations, while the remaining algorithms exploit QuIDD properties explicitly. The section concludes with experiments comparing all the algorithms.

Inner Product. Since the quantum-circuit formalism models an arbitrary quantum state $|\psi\rangle$ as a unit vector, the inner product $\langle\psi|\psi\rangle = 1$. In the case of a global-phase difference between two states $|\psi\rangle$ and $|\varphi\rangle$, the inner product is the global-phase factor, $\langle\varphi|\psi\rangle = e^{i\theta}\langle\psi|\psi\rangle = e^{i\theta}$. Since $|e^{i\theta}| = 1$ for any θ, checking if the complex modulus of the inner product is 1 suffices to check global-phase equivalence for states.

Although the inner product may be computed using explicit arrays, a straightforward QuIDD-based implementation is easily derived. The complex-conjugate transpose and matrix product with QuIDD operands were defined in Chapter 8. Thus, the algorithm computes the complex-conjugate transpose of A and multiplies the result with B. The complexity of this algorithm is stated in the following lemma.

Lemma 9.1. *Consider two state QuIDDs A and B with sizes $|A|$ and $|B|$, respectively, in number of nodes. The global-phase difference, if any, can be computed in $O(|A||B|)$ time and memory.*

Proof. Computing the complex-conjugate transpose of A requires $O(|A|)$ time and memory (Section 7.1). Matrix multiplication of two ADDs of sizes $|A|$ and $|B|$ requires $O((|A||B|)^2)$ time and memory (Section 7.1). However, this bound is loose for an inner product because only a single dot product must be performed. In this case, the ADD matrix multiplication algorithm reduces to a single call of $C = \textbf{Apply}(A, B, *)$ followed by $D = \textbf{Apply}(C, +)$ [6]. D is a single terminal node containing the global-phase factor if $|value(D)| = 1$. $\textbf{Apply}(A, B, *)$ and $\textbf{Apply}(C, +)$ are computed in $O(|A||B|)$ time and memory [20], while $|value(D)|$ is computed in $O(1)$ time and memory. □

Matrix Product. The matrix product of two operators can be used for global-phase equivalence checking. In particular, since all quantum operators are unitary, the adjoint of each operator is its inverse. Thus, if two operators U and V differ by a global phase, then $UV^\dagger = e^{i\theta}I$.

With QuIDD representations of U and V, computing $V^\dagger$ requires $O(|V|)$ time and memory (Section 7.1). The matrix product $W = UV^\dagger$ requires $O((|U||V|)^2)$ time and memory (Section 7.1). To check if $W = e^{i\theta}I$, any terminal value t is chosen from W, and scalar division is performed on W as $W' = \textbf{Apply}(W, t, /)$, which takes $O((|U||V|)^2)$ time and memory. Since QuIDDs are canonical, checking if $W' = I$ requires only $O(1)$ time and memory. If $W' = I$, then t is the global-phase factor.

Node-Count Check. The previous algorithms merely translate linear-algebraic operations to QuIDDs, but exploiting the following QuIDD property leads to faster checks.

Lemma 9.2. *The QuIDD* $A' = \textbf{Apply}(A, c, *)$, *where* $c \in \mathbb{C}$ *and* $c \neq 0$, *is isomorphic to* A, *hence* $|A'| = |A|$.

Proof. In creating A', **Apply** expands all of the internal nodes of A since c is a scalar, and the new terminals are the terminals of A multiplied by c. All terminal values t_i of A are unique by definition of a QuIDD (see Chapter 7). Thus, $ct_i \neq ct_j$ for all i, j such that $i \neq j$. As a result, the number of terminals in A' is the same as in A. □

Lemma 9.2 states that two QuIDD states or operators that differ by a non-zero scalar, such as a global-phase factor, have the same number of nodes. Thus, equal node counts in QuIDDs is a necessary but not sufficient condition for global-phase equivalence. To see why it is not sufficient, consider two state vectors $|\psi\rangle$ and $|\varphi\rangle$ with elements w_j and v_k, respectively, where $j, k = 0, 1, \ldots N-1$. If some $w_j = v_k = 0$ such that $j \neq k$, then $|\varphi\rangle \neq e^{i\theta}|\psi\rangle$. The QuIDD representations of these states can, in general, have the same node counts. Despite this drawback, the node-count check requires only $O(1)$ time since **Apply** is easily augmented to recursively sum the number of nodes as a QuIDD is created.

Recursive Check. Lemma 9.2 implies that a QuIDD-based algorithm which takes into account terminal value differences implements a sufficient condition for checking global-phase equivalence. The pseudocode for such an algorithm called **GPRC** is presented in Figure 9.2.

GPRC returns **true** if two QuIDDs A and B differ by global phase and **false** otherwise. gp and $have\_gp$ are global variables containing the global-phase factor and a flag signifying whether or not a terminal node has been reached, respectively. The value of gp is only valid if **true** is returned.

The first conditional block of **GPRC** deals with terminal values. The potential global-phase factor ngp is computed after handling division by 0. If $|ngp| \neq 1$ or if $ngp \neq gp$ when gp has been set, then the two QuIDDs do not differ by a global phase. Next, the condition specified by Lemma 9.2 is addressed. If the node of A depends on a different row or column variable than the node of B, then A and B are not isomorphic and thus cannot differ by global phase. Finally, **GPRC** is called recursively, and the results of these calls are combined via the logical AND operation.

```
GPRC(A, B, gp, have_gp) {
  if (Is_Constant(A) and Is_Constant(B)) {
    if (Value(B) == 0) return (Value(A) == 0)
    ngp = Value(A)/Value(B)
    if (sqrt(real(ngp) * real(ngp)+
      imag(ngp) * imag(ngp)) != 1)
        return false
    if (!have_gp) {
      gp = ngp
      have_gp = true;
    }
    return (ngp == gp)
  }
  if ((Is_Constant(A) and !Is_Constant(B))
    or (!Is_Constant(A) and Is_Constant(B)))
      return false;
  if (Var(A)! = Var(B)) return false
  return (GPRC(Then(A), Then(B), gp, have_gp)
   and GPRC(Else(A), Else(B), gp, have_gp))
}
```

Fig. 9.2. Recursive global-phase equivalence check.

Early termination occurs when isomorphism is violated or more than one phase difference is computed. In the worst case, both QuIDDs will be isomorphic, but the last terminal visited in each QuIDD will differ by more than a global-phase factor, causing full traversals of both QuIDDs. Thus, the overall runtime and memory complexity of **GPRC** for states or operators is $O(|A|+|B|)$. Also, the node-count check can be run before **GPRC** to quickly eliminate many nonequivalences.

Empirical Results. The first benchmark considered is a single iteration of Grover's quantum search algorithm [38], which is depicted in Figure 9.3. As in Chapter 7, the oracle searches for the last item in the database. One iteration is sufficient to test the effectiveness of the algorithms, since the state vector QuIDD remains isomorphic across all iterations, as proven in Section 7.2.

Figure 9.4a shows the runtime results for the inner product and **GPRC** algorithms (no results are given for the node-count check algorithm since it runs in $O(1)$ time). These results confirm the asymptotic complexity difference between the inner-product and **GPRC** algorithms. The number of nodes in the QuIDD state vector after a Grover iteration is $O(n)$ [99], which is confirmed in Figure 9.4b. As a result, the runtime complexity of the inner product should be $O(n^2)$, which is confirmed by a regression plot within 1% error. In contrast, the runtime complexity of the **GPRC** algorithm should be $O(n)$, which is also confirmed by another regression plot within the same error.

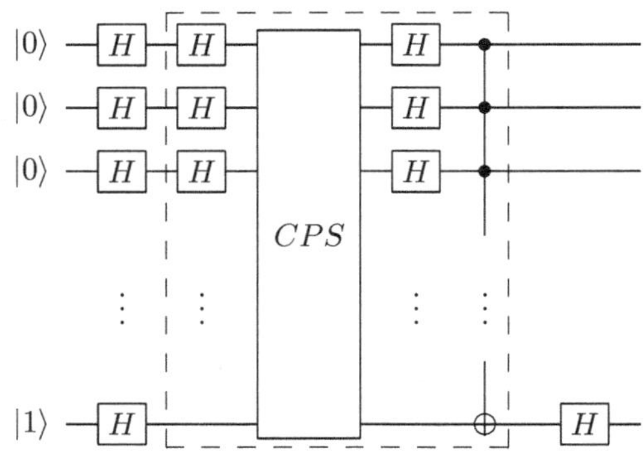

Fig. 9.3. One iteration of Grover's search algorithm with an ancillary qubit used by the oracle. CPS is the (conditional) phase shift operator and the dashed portion is the Grover iteration operator.

Figure 9.5a shows runtime results for the matrix product and **GPRC** algorithms checking the Grover operator. We showed in Chapter 7 that the QuIDD representation of this operator grows in size as $O(n)$, which is confirmed in Figure 9.5b. Therefore, the runtime of the matrix product should be quadratic in n, but linear in n for **GPRC**. Regression plots verify these complexities within 0.3% error.

The next benchmark compares states that appear in Shor's integer factorization algorithm [82]. In particular, we consider states created by the modular exponentiation sub-circuit that represent all possible combinations of x and $f(x, N) = a^x \ mod\ N$, where N is the integer to be factored (see Figure 9.6). Each of the $O(2^n)$ paths to a non-0 terminal encodes a binary value for x and $f(x, N)$. This experiment demonstrates how the algorithms fare with exponentially-growing QuIDDs.

Tables 9.1a-d show the results of the inner product and **GPRC** for this benchmark. Each N is an integer whose two non-trivial factors are prime.[2] a is set to $N-2$ since it may be chosen randomly from the range $[2..N-2]$. In the case of Table 9.1a, states $|\psi\rangle$ and $|\varphi\rangle$ are equal up to global phase. The node counts for both states are equal as predicted by Lemma 9.2. Interestingly, both algorithms exhibit nearly the same performance. Tables 9.1b-d contain results for the cases in which Hadamard gates are applied to the first, middle, and last qubits, respectively, of $|\varphi\rangle$. These results show that early termination in **GPRC** can enhance performance by a factor of roughly 1.5x to 10x.

In almost every case, both algorithms represent far less than 1% of the total runtime. Thus, checking for global-phase equivalence among QuIDD states appears to be an easily achievable task once the QuIDDs are created. An

[2] Such integers are likely to be the ones input to Shor's algorithm since they are the foundation of modern public key cryptography [82].

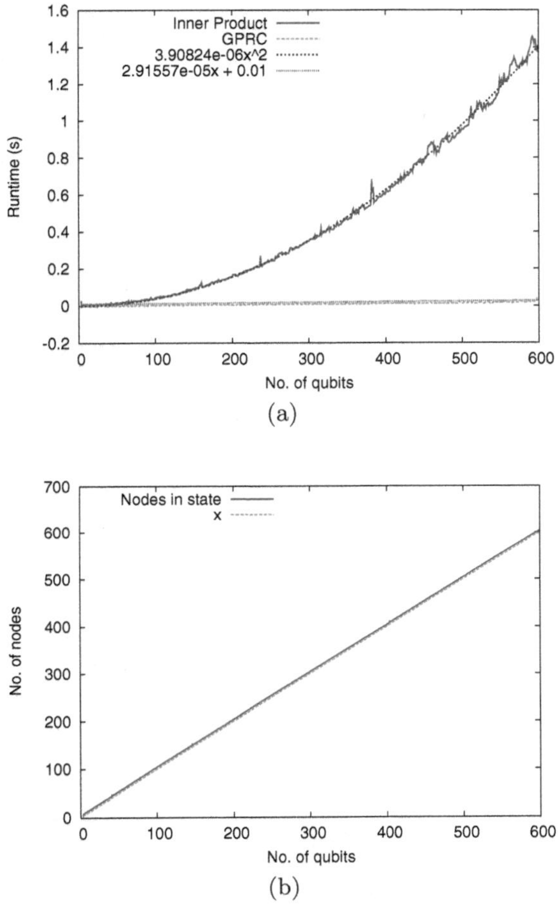

Fig. 9.4. (a) Runtime results for the inner product and **GPRC** on checking global-phase equivalence of states generated by a Grover iteration. (b) Node counts for the QuIDD representation of the state vector.

interesting side note is that in modular exponentiation, some QuIDD states with more qubits have more exploitable structure than those with fewer qubits. For instance, the $N = 387929$ (19 qubits) QuIDD has fewer than half the nodes of the $N = 163507$ (18 qubits) QuIDD.

Table 9.2 contains results for the matrix-product and **GPRC** algorithms checking the inverse QFT operator. As noted in Chapter 7, the inverse QFT is a key operator in Shor's algorithm [82], and its n-qubit QuIDD representation grows as $O(2^{2n})$. In this case, the asymptotic differences in the matrix product and **GPRC** are very noticeable. Also, the memory usage indicates that the matrix product may need asymptotically more intermediate memory despite operating on QuIDDs with the same number of nodes as **GPRC**.

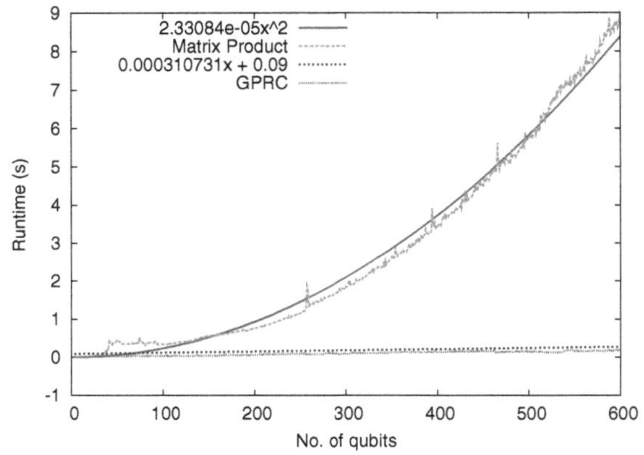

(a)

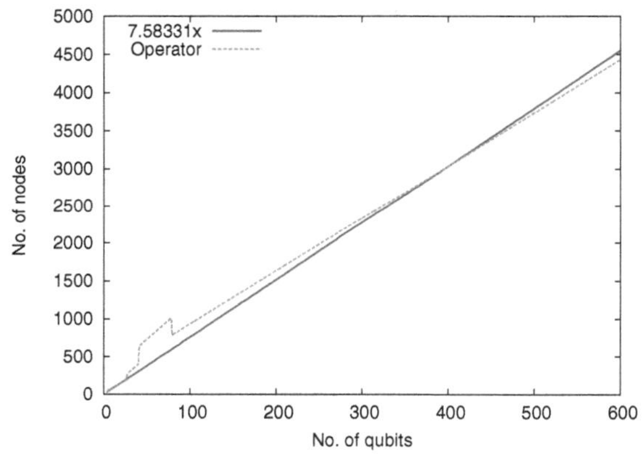

(b)

Fig. 9.5. (a) Runtime results for the matrix product and **GPRC** on checking global-phase equivalence of the Grover iteration operator. (b) Node counts for the QuIDD representation of the operator.

9.3 Relative-Phase Equivalence

Like the global-phase case, the relative-phase equivalence checking problem can be solved in several ways. The first three algorithms adapt standard linear algebra to QuIDDs, while the last two algorithms exploit QuIDD properties directly, offering asymptotic runtime and memory improvements.

Modulus and Inner Product. Consider two state vectors $|\psi\rangle$ and $|\varphi\rangle$ that are equal up to relative phase and have complex-valued elements w_j and v_k, respectively, where $j, k = 0, 1, \ldots, N-1$. Computing $|\varphi'\rangle = \Sigma_{i=0}^{N-1}|v_j|\,|j\rangle$ and $|\psi'\rangle = \Sigma_{k=0}^{N-1}|w_k|\,|k\rangle = \Sigma_{k=0}^{N-1}|e^{i\theta_k}v_k|\,|k\rangle$ sets each phase factor to 1, allowing the inner product to be applied as in Section 9.2. The complex modulus

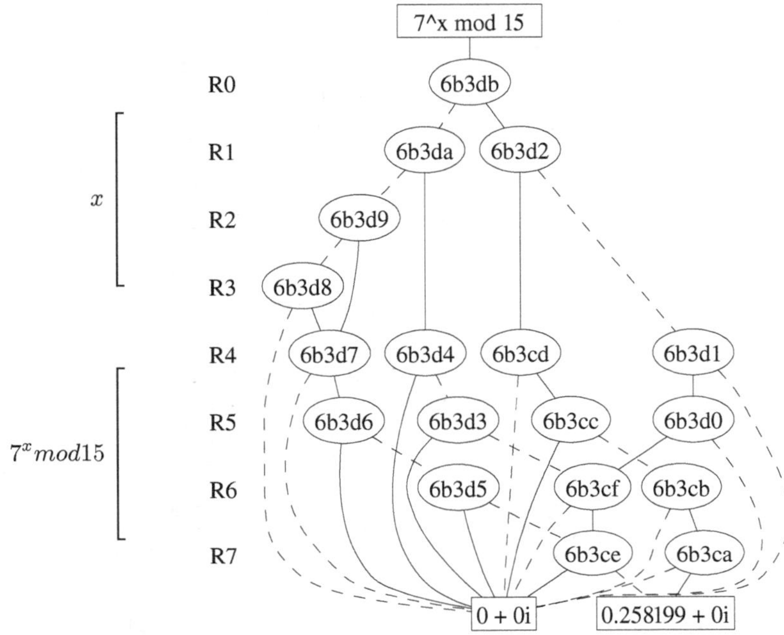

Fig. 9.6. A QuIDD state combining x and 7^x *mod* 15 in binary. The first qubit of each partition is least significant. Internal node labels are unique hexadecimal identifiers based on each node's memory address.

operations are computed as $C = \mathbf{Apply}(A, |\cdot|)$ and $D = \mathbf{Apply}(B, |\cdot|)$ with runtime and memory complexity $O(|A| + |B|)$, which is dominated by the $O(|A||B|)$ inner product complexity.

Modulus and Matrix Product. For operator equivalence up to relative phase, there are two cases depending on whether the diagonal relative-phase matrix appears on the left or right side of one of the operators. Consider two operators U and V with elements $u_{j,k}$ and $v_{j,k}$, respectively, where $j, k = 0, \ldots N-1$. The two cases in which the relative-phase factors appear on either side of V are described as $u_{j,k} = e^{i\theta_j} v_{j,k}$ (left side) and $u_{j,k} = e^{i\theta_k} v_{j,k}$ (right side). In either case, the matrix product check discussed in Section 9.2 may be extended by computing the complex modulus without increasing the overall complexity. Note that neither this algorithm nor the modulus and inner-product algorithms calculate the relative-phase factors.

Element-wise Division. Given the states discussed for the modulus and inner product check, $w_k = e^{i\theta_k} v_k$, the operation w_k/v_j for each $j = k$ is a relative-phase factor, $e^{i\theta_k}$. The condition $|w_k/v_j| = 1$ is used to check if each division yields a relative phase. If this condition is satisfied for all divisions, the states are equal up to relative phase.

No. of qubits	N	Creation time (s)	No. of nodes $\|\psi\rangle$	No. of nodes $\|\varphi\rangle$	IP time (s)	GPRC time (s)
10	993	2.37	273	273	0.012	0.008
11	1317	3.23	1710	1710	0.064	0.048
12	4031	11.9	9391	9391	0.30	0.26
13	6973	24.8	10680	10680	0.34	0.28
14	12127	55.1	18236	18236	0.54	0.46
15	19093	128.3	12766	12766	0.41	0.32
16	50501	934.1	51326	51326	1.7	1.6
17	69707	1969	26417	26417	0.87	0.78
18	163507	12788	458064	458064	19.6	19.6
19	387929	93547	182579	182579	6.62	6.02

(a)

No. of nodes $\|\varphi\rangle$	IP time (s)	GPRC time (s)
508	0.012	$< 1e-10$
1812	0.052	0.004
10969	0.27	0.036
11649	0.31	0.036
19978	0.54	0.06
13446	0.41	0.036
55447	1.53	0.2
27797	0.78	0.084
521725	19.0	9.18
194964	6.44	4.40

(b)

No. of qubits	N	Creation time (s)	No. of nodes $\|\psi\rangle$	No. of nodes $\|\varphi\rangle$	IP time (s)	GPRC time (s)
10	993	2.37	273	508	0.016	$< 1e-10$
11	1317	3.23	1710	2768	0.068	0.024
12	4031	11.9	9391	11773	0.27	0.076
13	6973	24.8	10680	16431	0.43	0.14
14	12127	55.1	18236	29584	0.65	0.22
15	19093	128.3	12766	19207	0.56	0.20
16	50501	934.1	51326	71062	1.76	0.84
17	69707	1969	26417	46942	1.24	0.55
18	163507	12788	458064	653048	31.7	26.1
19	387929	93547	182579	312626	9.33	6.44

(c)

No. of nodes $\|\varphi\rangle$	IP time (s)	GPRC time (s)
508	0.008	0.004
2768	0.056	0.008
14092	0.21	0.088
16431	0.27	0.084
29584	0.53	0.13
19207	0.50	0.084
74919	1.51	0.66
46942	1.13	0.25
629533	29.6	23.7
312626	13.0	8.62

(d)

Table 9.1. Performance results for the inner product and **GPRC** algorithms on checking global-phase equivalence of modular exponentiation states. In (a), $|\psi\rangle = |\varphi\rangle$ up to global phase. In (b), (c), and (d), Hadamard gates are applied to the first, middle, and last qubits of $|\varphi\rangle$ so that $|\psi\rangle \neq |\varphi\rangle$ up to global phase.

No. of qubits	Matrix product Time (s)	Matrix product Memory (MB)	GPRC Time (s)	GPRC Memory (MB)
3	0.036	0.13	0.004	0.13
4	0.30	0.39	0.016	0.13
5	2.53	1.41	0.064	0.25
6	22.55	6.90	0.24	0.66
7	271.62	46.14	0.98	2.03
8	3637.14	306.69	4.97	7.02
9	22717	1800.42	17.19	26.48
10	—	$> 2GB$	75.38	102.4
11	—	$> 2GB$	401.34	403.9

Table 9.2. Performance results for the matrix product and **GPRC** algorithms on checking global-phase equivalence of the QFT operator used in Shor's algorithm. "$> 2GB$" indicates that memory usage of 2GB was exceeded.

The QuIDD implementation for states is simply $C = \textbf{Apply}(A, B, /)$, where **Apply** is augmented to avoid division by 0 and instead return 1 when two terminal values being compared equal 0, and return 0 otherwise. **Apply** can be further augmented to terminate early when $|w_j/v_i| \neq 1$. C is a QuIDD vector containing the relative-phase factors. If C contains a terminal value of 0, then A and B do not differ by relative phase. Since a call to **Apply** implements this algorithm, the runtime and memory complexity are $O(|A||B|)$.

Element-wise division for operators is more complicated. For QuIDD operators U and V, W = **Apply**$(U, V, /)$ corresponds to a matrix with the relative-phase factor $e^{i\theta_j}$ along row j in the case of phases appearing on the left side, and along column j in the case of phases appearing on the right side. In the first case, all rows of W are identical, meaning that the support of W does not contain any row variables. Similarly, in the second case the support of W does not contain any column variables. A complication arises when 0 values appear in either operator. In such cases, the support of W may contain both variable types, but the operators may in fact be equal up to relative phase. Figure 9.11 presents an algorithm based on **Apply**, which accounts for these special cases by using a sentinel value of 2 to mark valid 0 entries that do not affect relative-phase equivalence.[3] These entries are recursively ignored by skipping either row or column variables with sentinel children (S specifies row or column variables), which effectively fills copies of neighboring row or column phase values in their place in W. The algorithm must be run twice, once for each variable type. The size of W is $O(|U||V|)$ since it is created by a variant of **Apply**.

Non-0 Terminal Merge. A necessary condition for relative-phase equivalence is that zero-valued elements of each state vector appear in the same locations, as expressed by the following lemma.

Lemma 9.3. *A necessary but not sufficient condition for two states* $|\varphi\rangle = \Sigma_{j=0}^{N-1} v_j |j\rangle$ *and* $|\psi\rangle = \Sigma_{k=0}^{N-1} w_k |k\rangle$ *to be relative-phase equivalent is that* $\forall v_j = w_k = 0,\ j = k$.

Proof. If $|\psi\rangle = |\varphi\rangle$ up to relative phase, $|\psi\rangle = \Sigma_{k=0}^{N-1} e^{i\theta_k} v_k |k\rangle$. Since $e^{i\theta_k} \neq 0$ for any θ, if any $w_k = 0$, then $v_j = 0$ must also be true where $j = k$. A counter-example proving insufficiency is $|\psi\rangle = (0, 1/\sqrt{3}, 1/\sqrt{3}, 1/\sqrt{3})^T$ and $|\varphi\rangle = (0, 1/2, 1/\sqrt{2}, 1/2)^T$. □

QuIDD canonicity may be exploited with this condition. Let A and B be the QuIDD representations of the states $|\psi\rangle$ and $|\varphi\rangle$, respectively. First compute C = **Apply**$(A, \lceil|\cdot|\rceil)$ and D = **Apply**$(B, \lceil|\cdot|\rceil)$, which converts every non-zero terminal value of A and B into a 1. Since C and D have only two terminal values, 0 and 1, checking if C and D are equal satisfies Lemma 9.3. Canonicity ensures this check requires $O(1)$ time and memory. The overall runtime and memory complexity of this algorithm is $O(|A| + |B|)$ due to the unary **Apply** operations. This algorithm can also be applied to operators since Lemma 9.3 also applies to $u_{j,k} = e^{i\theta_j} v_{j,k}$ (phases on the left) and $u_{j,k} = e^{i\theta_k} v_{j,k}$ (phases on the right) for operators U and V.

[3] Sentinels are special terminal entries. They do not contain mere 0 values, but rather *valid* 0 values that do not affect equivalence and should therefore be treated differently by the algorithm. Any sentinel value larger than 1 may be used since such values do not appear in the context of quantum circuits.

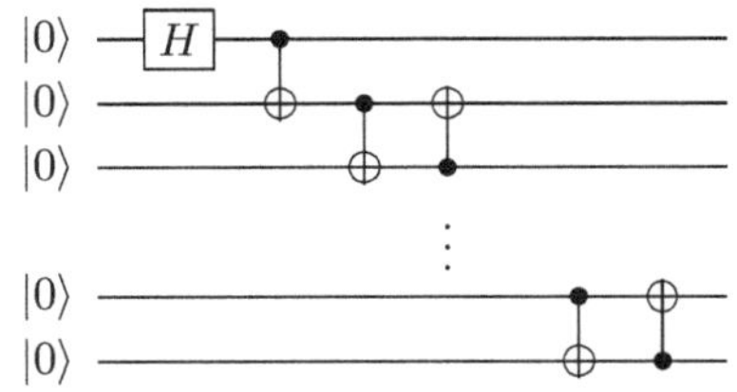

Fig. 9.7. Remote EPR-pair creation between the first and last qubits via nearest-neighbor interactions.

Modulus and DD Compare. A variant of the modulus and inner-product check, which also exploits the canonicity of QuIDDs, provides an asymptotic improvement when checking a necessary and sufficient condition for relative-phase equivalence of states and operators. As with the modulus and inner product check, compute $C = \mathbf{Apply}(A, |\cdot|)$ and $D = \mathbf{Apply}(B, |\cdot|)$. If A and B are equal up to relative phase, then $C = D$ since each phase factor becomes a 1. Canonicity again ensures this check requires $O(1)$ time and memory. Thus, the runtime and memory complexity of this algorithm is dominated by the unary **Apply** operations, giving $O(|A| + |B|)$.

9.4 Empirical Validation

We now present experimental results for the relative-phase equivalence-checking algorithms. The first benchmark circuit creates a remote EPR pair between the first and last qubits), via nearest-neighbor interactions [13]. The circuit is shown in Figure 9.7 and is discussed in detail later in Chapter 10. Given an initial state $|00\ldots0\rangle$, it creates the remote EPR-pair state $(1/\sqrt{2})(|00\ldots0\rangle + |10\ldots1\rangle)$. The circuit size is varied, and the final state is compared to the state $(e^{0.345i}/\sqrt{2})\,|00\ldots0\rangle + (e^{0.457i}/\sqrt{2})\,|10\ldots1\rangle$.

Runtime results for all algorithms are provided in Figure 9.8a. They show that all the algorithms run quickly. For example, the inner product is the slowest algorithm, yet for a 1000-qubit instance, it runs in approximately 0.2 seconds, a small fraction of the 7.6 seconds required to create the QuIDD state vectors.

Regressions of the runtime and memory data reveal linear complexity for all algorithms to within 1% error. This is not unexpected since the QuIDD representations of the states grow linearly with the number of qubits (see Figure 9.8b), and the complex modulus reduces the number of different terminals prior to computing the inner product. These results suggest that in practice, the inner-product and element-wise division algorithms perform better than their worst-case complexity. Element-wise division should be preferred when

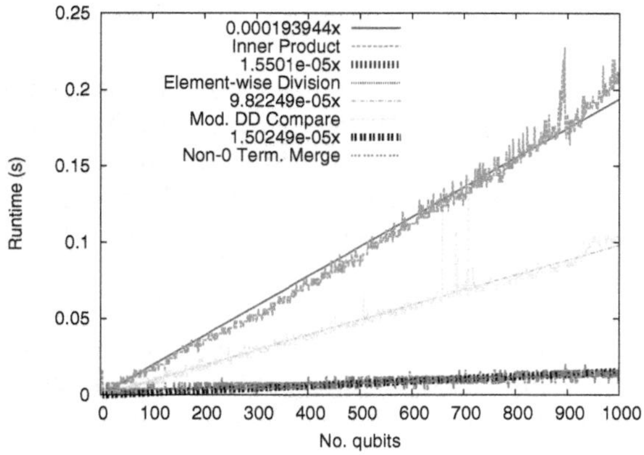

(a)

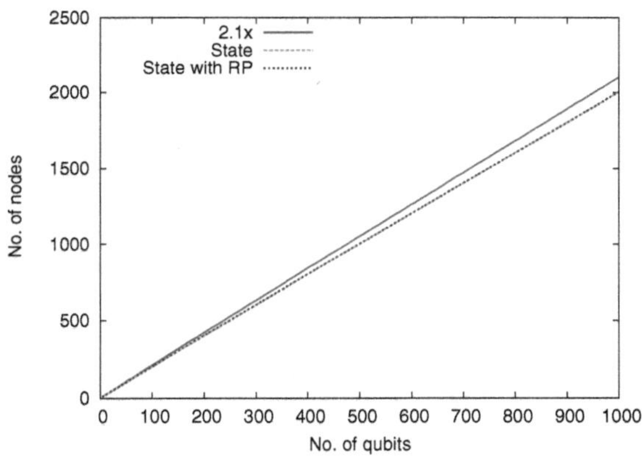

(b)

Fig. 9.8. (a) Runtime results for the inner product, element-wise division, modulus and DD compare, and non-0 terminal merge algorithms for checking relative-phase equivalence of the remote EPR pair circuit. (b) Node count as a function of the number of qubits.

QuIDD states are compact since, unlike the other algorithms, it computes the relative-phase factors.

The Hamiltonian simulation circuit shown in Figure 9.9 is taken from [61, Figure 4.19, p. 210]. When its one-qubit gate (boxed) varies with Δt, it produces a variety of diagonal operators, all of which are equivalent up to relative phase. Empirical results for such equivalence checking are shown in Figure 9.10. As in the case of the teleportation circuit benchmark, the matrix product and element-wise division algorithms perform better than their worst-case asymptotic upper bounds, indicating that element-wise division is the best choice for compact QuIDD operators.

9.5 Summary

Although QuIDD properties like canonicity enable exact equivalence checking in $O(1)$ time, we have shown that such properties may be further exploited to develop efficient algorithms for the difficult problem of equivalence checking up to global and relative phase. In particular, the global-phase recursive check and element-wise division algorithms efficiently determine equivalence of states and operators up to global and relative phase, while also computing the phases. In practice, they outperform QuIDD implementations of the inner and matrix product, which do not compute relative-phase factors. Other QuIDD algorithms presented in this chapter, such as the node-count check, non-0 terminal merge, and modulus and DD compare, further exploit decision-diagram properties to provide even faster checks, but only satisfy necessary conditions for equivalence. Thus, they should be used to aid the more robust algorithms. A summary of the theoretical results presented here appears in Table 9.3.

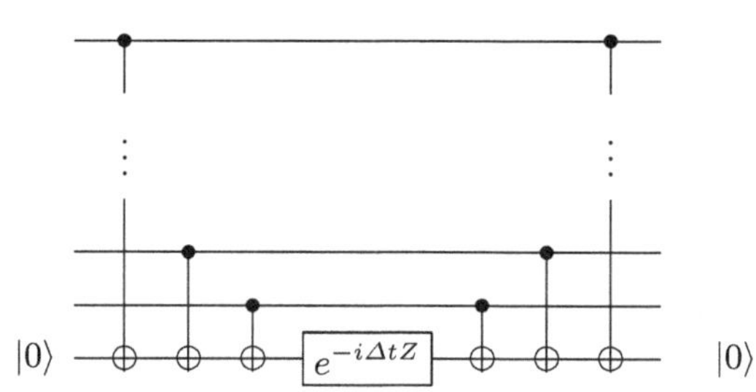

Fig. 9.9. Quantum-circuit realization of a Hamiltonian consisting of Pauli operators. Extra Pauli gates may be needed for some Hamiltonians.

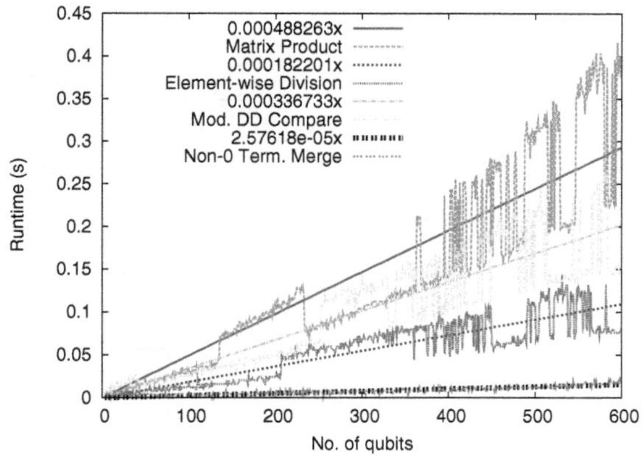

(a)

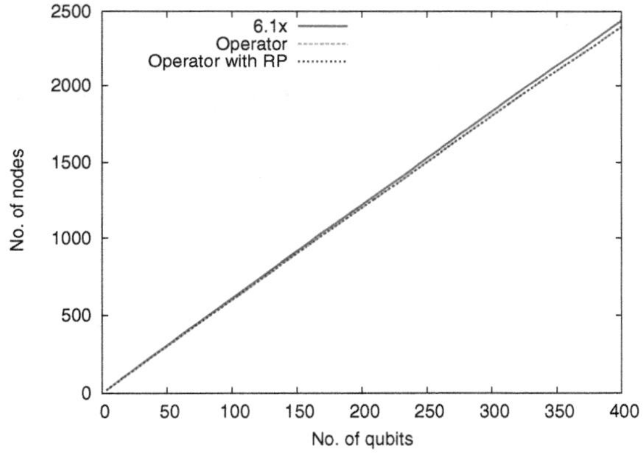

(b)

Fig. 9.10. (a) Runtime results for the matrix product, element-wise division, modulus and DD compare, and non-0 terminal merge algorithms to check for relative-phase equivalence in the Hamiltonian Δt circuit. (b) Node count as a function of the number of qubits.

Algorithm	Phase type	Finds phases?	Necessary & sufficient?	$O(\cdot)$ time complexity: best-case	$O(\cdot)$ time complexity: worst-case
Inner product	Global	Yes	Yes	$\|A\|\|B\|$	$\|A\|\|B\|$
Matrix product	Global	Yes	Yes	$(\|A\|\|B\|)^2$	$(\|A\|\|B\|)^2$
Node-count	Global	No	Nec. only	1	1
Recursive check	Global	Yes	Yes	1	$\|A\| + \|B\|$
Modulus and inner product	Relative	No	Yes	$\|A\|\|B\|$	$\|A\|\|B\|$
Element-wise division	Relative	Yes	Yes	$\|A\|\|B\|$	$\|A\|\|B\|$
Non-0 terminal merge	Relative	No	Nec. only	$\|A\| + \|B\|$	$\|A\| + \|B\|$
Modulus and DD compare	Relative	No	Yes	$\|A\| + \|B\|$	$\|A\| + \|B\|$

Table 9.3. Key properties of the QuIDD equivalence-checking algorithms.

```
RP_DIV(A, B, S) {
  if (A == New_Terminal(0)) {
    if (B != New_Terminal(0))
      return New_Terminal(0)
    return New_Terminal(2)
  }
  if (Is_Constant(A) and Is_Constant(B)) {
    nrp = Value(A)/Value(B)
    if (sqrt(real(nrp) * real(nrp)+
      imag(nrp) * imag(nrp)) != 1)
        return New_Terminal(0)
    return New_Terminal(nrp)
  }
  if (Table_Lookup(R, RP_DIV, A, B, S)) return R;
  v = Top_Var(A, B)
  T = RP_DIV(A_v, B_v, S)
  E = RP_DIV(A_v', B_v', S)
  if ((T == New_Terminal(0)) or
    (E == New_Terminal(0)))
      return New_Terminal(0)
  if ((T != E) and (Type(v) == S)) {
    if (Is_Constant(T) and Value(T) == 2)
      return E
    if (Is_Constant(E) and Value(E) == 2)
      return T
    return New_Terminal(0)
  }
  if (Is_Constant(T) and Value(T) == 2)
    T = New_Terminal(1)
  if (Is_Constant(E) and Value(E) == 2)
    E = New_Terminal(1)
  R = ITE(v, T, E)
  Table_Insert(R, RP_DIV, A, B, S)
  return R
}
```

Fig. 9.11. Element-wise division algorithm.

10

Improving QuIDD-based Simulation

This chapter describes some ways to speed up QuIDD-based simulation that are facilitated by the QuIDDPro language. These techniques apply to QuIDD matrix multiplication, the tensor product, and the partial trace, which are key operations for simulation with and without errors. First, we describe algorithms for efficiently applying controlled- and 1-qubit gates to QuIDD state vectors. The simulator uses these algorithms when processing particular source-code expressions in its input. Next, we show how one can selectively tensor and remove, via the partial trace operation, qubits that do not affect a computation's final outcome. Although this technique cannot be used in every case, it can exponentially reduce simulation complexity in important special cases. We illustrate this by simulating a circuit with several types of continuous errors and evaluating the effectiveness of "bang-bang" error correction.

10.1 Gate Algorithms

Recall that matrix-vector and matrix-matrix multiplication are the main operations for simulating quantum circuits, as they provide the mathematical machinery for applying gates to qubits. QuIDDs utilize a variant of the ADD-based matrix multiplication algorithm described in Section 7.1. This algorithm is a fairly straightforward translation of dot products to the graph domain and makes use of the **Apply** procedure. It assumes that the two QuIDD arguments have the same dimension. A consequence of this assumption is that for small gates of, say, 1- and 2-qubits, a larger operator must be constructed by computing tensor products with identity matrices. For example, to apply a 1-qubit gate to some qubit of a 5-qubit state vector or density matrix, the matrix representing the gate must be first tensored with four 1-qubit identity matrices and then multiplied with the QuIDD representing the entire state vector or density matrix. It is natural to ask: Can a more clever approach be used to apply gates by leveraging the peculiar properties of QuIDDs, at least for certain types of gates? Such an improvement is indeed possible, as

$$UV = \begin{bmatrix} u_{00} & u_{01} \\ u_{10} & u_{11} \end{bmatrix} \begin{bmatrix} v_0 \\ v_1 \end{bmatrix} = \begin{bmatrix} u_{00}v_0 + u_{01}v_1 \\ u_{10}v_0 + u_{11}v_1 \end{bmatrix}$$

(a)

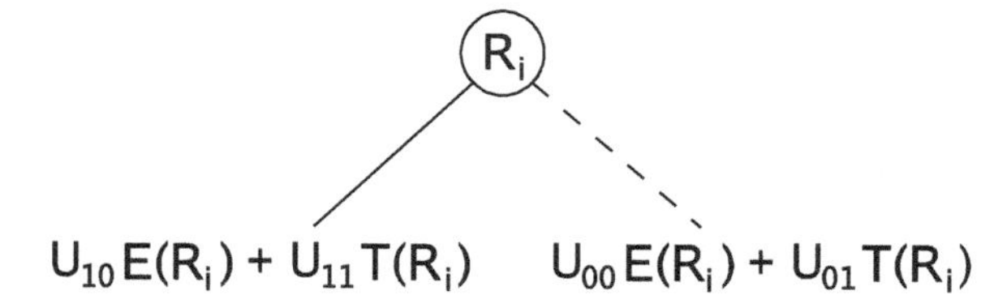

(b)

Fig. 10.1. (a) A 1-qubit gate applied to a single qubit, and (b) the QuIDD state vector transformation induced by this operation on qubit i.

we now explain. The specialized algorithms are described first, followed by a brief discussion of the relevant QuIDDPro language features. Furthermore, our QuIDDPro simulator automatically detects when such optimizations are applicable.

Simulating 1-qubit Gates

Special processing for small gates can be of great practical value considering that CNOT and all 1-qubit gates form a universal gate set [7]. The benefit of special processing for 1- and 2-qubit gates has been recognized previously, and is, in fact, the key notion underlying qubit-wise multiplication (Section 6.1). Unfortunately, qubit-wise multiplication only reduces the computational complexity of representing the operator and leaves the state vector or density matrix in an explicit, exponentially-sized form. A straightforward translation of the qubit-wise multiplication algorithm to QuIDDs would still result in exponential runtime since the algorithm iterates over all the indices of an array containing the state information.

What is needed is an algorithm that can both represent small operators concisely *and* update the state efficiently. An important property of QuIDDs is that each internal node R_i of a QuIDD state vector maps directly to qubit i since R_i represents the ith binary index of a state vector (Section 7.1). This means that applying a 1-qubit gate to qubit i can be accomplished simply by manipulating any instances of R_i nodes in a QuIDD state vector.

Given a QuIDD state vector, a top-down traversal is performed which transforms any R_i visited as shown in Figure 10.1. The transformation on R_i comes from the linear-algebraic description of a 1-qubit gate U acting on a 1-qubit state vector V to produce a new state vector V'. The probability amplitude for the $|0\rangle$ component of V' is $u_{00}v_0 + u_{01}v_1$. As a result, the subgraph pointed to by the 0 edge of the R_i node, or $E(R_i)$, is transformed into $u_{00}E(R_i)+u_{01}T(R_i)$. This operation is easily accomplished by two scalar multiplications via $E' = \textbf{Apply}(E(R_i), u_{00}, *)$ and $T' = \textbf{Apply}(T(R_i), u_{01}, *)$, followed by a single call to $\textbf{Apply}(E', T', +)$ to add the two results. The same

```
Q1_ALG(A, Op, qubit_index) {
  if (Table_Lookup(R, Q1_ALG, A, Op, qubit_index))
    return R
  v = Var(A)
  if (Is_Constant(A) or Index(v) >= qubit_index) {
    if (Index(v) == qubit_index)) {
      T = A_v
      E = A_v'
    }
    else T = E = A
    E00 = Apply(E, New_Terminal(Op_0,0), *)
    T01 = Apply(T, New_Terminal(Op_0,1), *)
    E10 = Apply(E, New_Terminal(Op_1,0), *)
    T11 = Apply(T, New_Terminal(Op_1,1), *)
    E = Apply(E00, T01, +)
    T = Apply(E10, T11, +)
    R = ITE(v, T, E)
    Table_Insert(R, Q1_ALG, A, Op, qubit_index)
    return R
  }
  T = Q1_ALG(A_v, Op, qubit_index)
  E = Q1_ALG(A_v', Op, qubit_index)
  R = ITE(v, T, E)
  Table_Insert(R, Q1_ALG, A, Op, qubit_index)
  return R
}
```

Fig. 10.2. Pseudocode for the 1-qubit gate algorithm. $\mathtt{Op}_{\mathtt{i,j}}$ denotes accessing the complex value at row i and column j of the 1-qubit matrix Op.

transformation is also performed on the subgraph pointed to by the 1 edge, except that u_{10} and u_{11} are used. If an R_i variable is missing in any particular path, which can be detected by encountering an R_j node such that $i < j$, then a new R_i node is created with children $u_{00}R_j + u_{01}R_j$ and $u_{10}R_j + u_{11}R_j$. Special checks on the node cache are performed to detect if the new children are equal, which results in the elimination of the R_i node as per the standard BDD rules. By performing this specialized 1-qubit gate operation on all R_i nodes in the QuIDD, the 1-qubit gate acting on qubit i need never be expanded into a larger n-qubit gate. All the extra overhead of the general ADD matrix multiplication algorithm is also avoided. Pseudocode for this algorithm is provided in Figure 10.2.

Simulating Controlled Gates

A controlled-U gate can also be implemented more efficiently using the 1-qubit gate QuIDD algorithm. Suppose a controlled-U gate is applied with qubit i as the control and qubit j as the target, such that $i < j$. As before, a top-down traversal is performed, but when any R_i node is encountered, the traversal only continues down the 1 edge of the R_i node, or $T(R_i)$, since a standard controlled-U gate performs no action when the control qubit is a $|0\rangle$. After proceeding down the 1 edge of any R_i node, U is applied on encountering any R_j node using the 1-qubit gate QuIDD algorithm. This operation is analogous to classical digital circuit simulation where the "controlling" values of logic gates are checked first before any other inputs [40]. For example, if an input wire of a k-input OR gate carries a 1 signal, then there is no need to check the other inputs since the output must be 1.

Interestingly, the controlled-U QuIDD algorithm is computationally more efficient than the 1-qubit gate QuIDD algorithm. This is simply because the controlled-U algorithm reduces the number of nodes in the QuIDD state vector that must be traversed by only traversing the 1 edges of controlling R_i nodes corresponding to control qubit i. Generalizing this result to controlled-U gates with multiple controls $i, i{+}1, \ldots, i{+}k$ such that $i{+}k < j$ shows that increasing the number of controls also increases the computational efficiency as each control further reduces the number of nodes to be traversed in the QuIDD state vector.

A further improvement can be made in the specific case of the CNOT gate. The top-down traversal proceeds as before, but only down the 1 edges of R_i nodes. However, when an R_j node is reached, the $E(R_j)$ and $T(R_j)$ subgraphs are simply swapped instead of applying the NOT gate. This can be done because the action of a NOT gate is precisely to switch the amplitudes of the $|0\rangle$ and $|1\rangle$ components of a qubit. A simple example of this algorithm operating on the QuIDD state vector $|11\rangle$ is depicted in Figure 10.3.

An important point to note is that the specialized controlled-U QuIDD algorithm only considers the case in which the control qubit precedes the target qubit. The reason that a bottom-up traversal cannot be used to implement a controlled-U gate whose target may precede one or more controls is due to the sharing of nodes across QuIDDs. For any DD, nodes are shared within the DD and *across* multiple instances of such data structures. This sharing across DDs not only increases efficiency, but it's a requirement for proper functioning since efficient construction of any new DD through the **Apply** function requires accessing the same node cache used by the DD arguments to **Apply** [20, 83]. As a result, there is no way for a bottom-up traversal to determine which DD it is in, since the terminal it starts at, and any subsequent node it visits, can be shared by multiple DDs. In contrast, a top-down traversal starts at the head of a specific DD.

If this is the case, then what can be done when a CNOT gate is applied to qubits i and j such that $i > j$? In this situation, the circuit equivalence for the

$$CNOT \begin{bmatrix} 0 \\ 0 \\ 0 \\ 1 \end{bmatrix} = \begin{bmatrix} 0 \\ 0 \\ 1 \\ 0 \end{bmatrix}$$

(a)

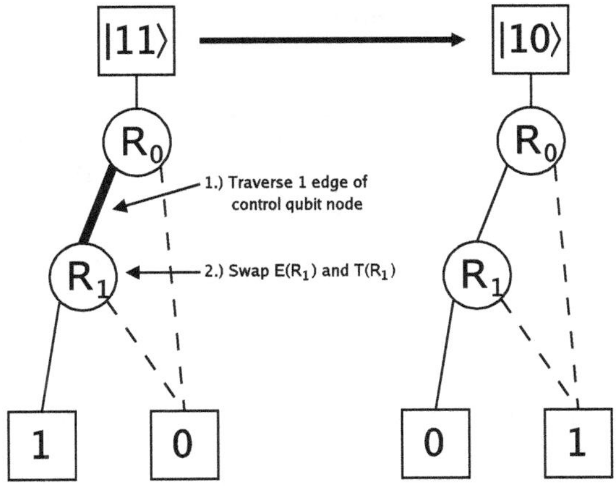

(b)

Fig. 10.3. (a) A CNOT gate applied to the $|11\rangle$ state vector, and (b) the same operation applied using the specialized QuIDD algorithm.

"upside-down" CNOT is employed as shown in Figure 10.4a [61]. Using this equivalence, 1-qubit Hadamards can be applied using the specialized 1-qubit algorithm in conjunction with the specialized CNOT algorithm. This means that applying specialized controlled-U algorithms to QuIDDs is computationally more efficient when the control qubit precedes the target qubit.

For the general case in which multiple controls are preceded by the target, swap gates [61] can be employed to swap the target qubit with the last control qubit. As shown in Figures 10.4a and 10.4b, the swap operation can be implemented with CNOT and 1-qubit Hadamard gates. In QuIDDPro, however, a special DD function is used to swap nodes. This function exhibits better performance by a large constant factor compared to applying an actual swap gate, since only one traversal of the QuIDD is needed.

Automatic Invocation of Relevant Algorithms

In order for these algorithms to be used without putting the burden on the user of detecting the special cases, the simulator should know when to apply them. To this end, we introduced a new function to the QuIDDPro input language (see Appendix A) called *cu_gate*. This function uses a string to specify which qubits are controls (or negated controls) and targets. To illustrate, suppose the user wants to apply a 1-qubit Pauli gate Y to qubits 4 and 7 conditional on control qubits 2 and 3, and negated control qubit 5. Further suppose that the total size of the circuit is 8 qubits. The *cu_gate* expression for this case is $cu\_gate(sigma\_y(1), \text{“}c2c3x4n5x7\text{''}, 8)$, where c, n, and x flag the subsequent qubit number as a control, negated control, or target, respectively. All unspecified qubits are assumed to be unaffected by the gate. Ordering within the specification string is irrelevant and handled internally by the simulator.

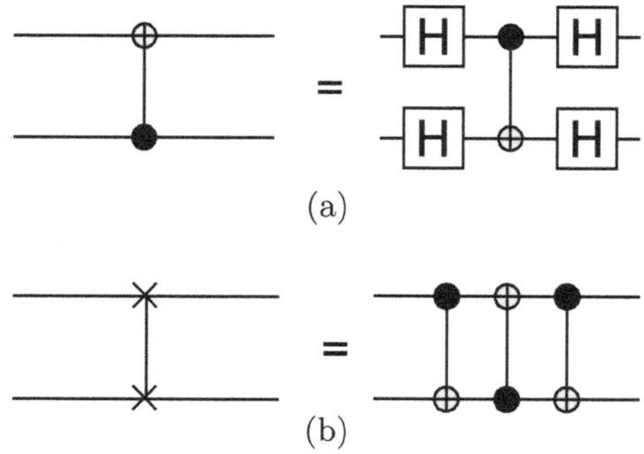

Fig. 10.4. (a) A CNOT whose target precedes its control is shown next to an equivalent circuit composed of 1-qubit Hadamard gates and a CNOT with the control and target qubits reversed. (b) A swap gate, which exchanges the values of two qubits, shown next to an equivalent circuit composed of CNOT gates. The CNOT gate in the center can be converted as shown in (a).

In the absence of the specialized gate-specific functions, QuIDDPro will create a QuIDD matrix according to the specifications given to *cu_gate*. This is accomplished efficiently with a series of tensor products and projections, all implemented with QuIDD algorithms. However, in the case of $cu\_gate(\cdot) * |state :>$, where $|state :>$ is a state vector QuIDD, the simulator does not create the QuIDD matrix. Instead, the specialized controlled-gate algorithm is performed on the state vector QuIDD directly. The specialized 1-qubit algorithm is applied when only "x's" exist in the string specification (1-qubit gates are viewed as controlled gates with no controls).

Unfortunately, since QuIDDPro has a very expressive input language[1], the situation becomes more complicated when the result of a call to *cu_gate* is stored in a variable and applied at a later time to a state vector QuIDD. For example, consider $U = cu\_gate(\cdot)$ followed arbitrarily later in the QuIDDPro script by $U * |state :>$. To handle such cases, operators created with *cu_gate* are *lazily evaluated*. In other words, QuIDDPro associates matrix variables with control/target information and no QuIDD matrix is created for as long as possible. When the gate is multiplied with another gate or printed to standard output, the QuIDD matrix is created only at that point in the simulation. As a result, if gates are always applied to state vectors, any gates created with $cu\_gate(\cdot)$ are never actually implemented, and the faster algorithms described in the previous sections are used instead. As we demonstrate next, this feature greatly enhances QuIDDPro's performance.

[1] QuIDDPro has approximately 100 built-in functions and other language features, which are detailed in Appendix A.

Empirical Results

We tested these specialized algorithms in QuIDDPro against the explicit ADD-based matrix multiplication algorithm. As evidenced by the results in Table 10.1, the specialized algorithms far outperform the matrix multiplication algorithm. "chp100" is a randomly generated 100-qubit circuit consisting of CNOT, Hadamard, and phase gates, which are the Clifford group generators (Chapter 5). "tchp100" is also a randomly generated 100-qubit circuit composed of Clifford group generators, but it also includes Toffoli gates. The addition of Toffoli gates is interesting since it forms a universal gate set for classical logic circuits [61]. "cnot200" stress-tests the specialized controlled gate algorithm since it is a randomly generated 200-qubit circuit consisting only of CNOT gates. Similarly, "toff200" is a circuit of the same size but with Toffoli gates only. As evidenced by the results, the performance improvements are as large as 60×. This indicates that the overhead avoided by specialization is significant. The results also demonstrate that the specialized algorithms allow QuIDDPro to simulate stabilizer circuits with and without (non-stabilizer) Toffoli gates very efficiently for large circuit sizes, making QuIDDPro competitive in practice with the stabilizer formalism.

			Specialized algorithms			ADD-based multiplication		
Benchmark	No. of qubits	No. of gates	Runtime (s)	Ave. time per gate (s)	Memory (MB)	Runtime (s)	Ave. time per gate (s)	Memory (MB)
chp100	100	300	4.57	0.0152	9.85	48.9	0.163	5.18
tchp100	100	300	0.870	0.00290	4.51	10.8	0.0361	1.61
cnot200	200	1000	2.54	0.00254	7.14	125	0.125	6.93
toff200	200	1000	4.61	0.00461	7.20	154	0.154	9.30

Table 10.1. Performance results comparing QuIDDPro using the specialized algorithms to QuIDDPro using ADD-based matrix multiplication.

10.2 Dynamic Tensor Products and Partial Tracing

This section discusses other language features related to the density matrix model which enable QuIDDPro to efficiently simulate a particular circuit of interest in the presence of continuous, random errors. Normally the size of a QuIDD is sensitive to the number of different matrix elements (see Chapter 7), but clever use of QuIDDPro's input language can reduce the negative effects of this sensitivity in certain cases. In particular, we dynamically add qubits to a density matrix state via the tensor product and removing them when they no longer affect the simulation results by tracing over them. The benchmark circuit, the error model used, and empirical results are also discussed in this section. The results include characterizations of imperfect gate errors, systematic errors, decoherence, and "bang-bang" error correction.

Language Support

In general, when some density state ρ_1 is not entangled with another state ρ_2, then $\rho_1 = tr_{\rho_2}(\rho_1 \otimes \rho_2)$ and similarly $\rho_2 = tr_{\rho_1}(\rho_1 \otimes \rho_2)$. In the course of simulation, if the qubits described by ρ_2 no longer affect the final states of interest, then they may be traced out to reduce simulation complexity. Rather than hold on to the separated state ρ_2 as in p-blocked simulation, it may be discarded entirely.

As shown below, circuits with nearest-neighbor interactions tend to contain qubits that have only a fleeting effect on the qubits of interest. The goal here is to introduce such qubits to the state only at the moment they are needed and to eliminate them from the state (and from memory) the moment they are no longer needed.

Since the QuIDDPro language provides a linear-algebraic interface, QuIDD matrices representing qubit states may be tensored at any time with density states of new qubits (see Appendix A). Furthermore, the partial trace may be efficiently performed at any time on any desired qubit (see Chapter 8). Although this technique is not automated, these language features allow the user to very easily implement the optimization at various points in circuits whose functionality is well-understood. We now describe one such circuit which we use as a case study to show the effectiveness of the proposed technique.

Experiments with Error Characterization

Some error-correcting codes have been developed to cope with errors in quantum hardware [87, 23, 34], but they require extra qubits and are most effective in the presence of single qubit errors only. Since the addition of extra qubits can be a daunting technological task, it can be very helpful to know a priori if error effects will be significant enough to require such correction. A different error correction approach has been proposed which involves applying corrective "$2\pi k$" pulses without the need for additional qubits or the single qubit error constraint [11]. The effectiveness of this technique has been demonstrated for teleporting qubits via remote entanglement. This refers to any entanglement between qubits that are not physical neighbors, and is achieved by a chain of nearest-neighbor interactions [13].

Although the aforementioned work specifically considers nuclear-spin quantum computing, remote entanglement through nearest-neighbor interaction is a common in other quantum computing technologies [111, 50, 66, 90, 18]. In the case of ion traps, even though qubits can be physically moved around, once in place, qubit interactions are performed between neighboring ions [47]. Equally important is the development of bang-bang error correction techniques which are a generalization of corrective pulses that decouple qubits from the environment, delaying the negative effects of decoherence errors for any technology [109, 52, 108, 45, 68]. As a result, simulating error effects in remote entanglement achieved by nearest-neighbor interactions using the technology-independent quantum circuit model forms an appealing case study.

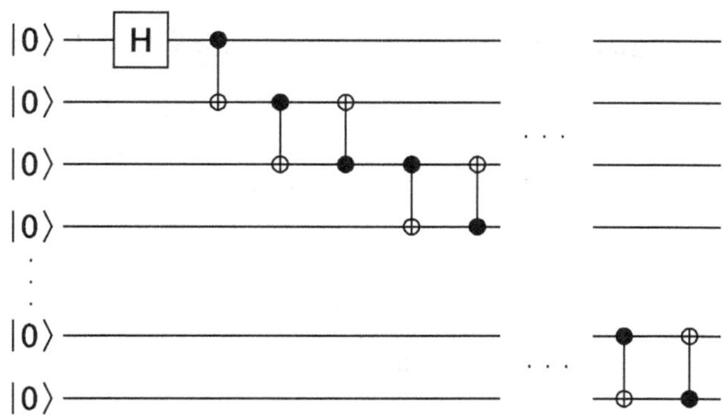

Fig. 10.5. Remote EPR-pair generation circuit which creates an EPR pair between qubits 0 (the top qubit) and $n-1$ (the bottom qubit) via nearest-neighbor interactions.

Remote Entanglement Circuits

Remote entanglement enables teleportation of an arbitrary quantum state from one party to another. The key ingredient in this scheme is the creation of an EPR pair between two communicating parties, Alice and Bob, as discussed in Section 3.2. Recall that the utility of this state lies in the fact that if Alice measures her particle and obtains a $|0\rangle$, then Bob will subsequently also obtain a $|0\rangle$ upon measurement of his particle. With only two qubits in the ground state, an EPR pair can be created by applying a Hadamard gate followed by a CNOT gate.

$$\Psi = (CNOT)(H \otimes I)\,|00\rangle = \frac{1}{\sqrt{2}}(|00\rangle + |11\rangle). \qquad (10.1)$$

In a circuit with $n > 2$ qubits, the above procedure can be generalized using nearest-neighbor interactions to create an EPR pair between qubits 0 and $n-1$ only. One straightforward generalization is to use a known nearest-neighbor decomposition of a CNOT gate with qubit 0 as the control and qubit $n-1$ as the target. Such a CNOT gate can be decomposed into $4(n-1)$ nearest-neighbor CNOT gates [76, Figure 3]. However, by making use of the fact that all qubits are initialized to the ground state, a smaller decomposition can be achieved with only $2n-3$ CNOT gates [13]. This circuit, which contains $2n-2$ gates, is shown in Figure 10.5. It generates the state

$$\Psi_R = \frac{1}{\sqrt{2}}(|00\ldots0\rangle + |10\ldots1\rangle).$$

In this state, qubits 0 and $n-1$ are remotely entangled since the measurement outcome of one qubit affects the measurement outcome of the other, yet the qubits are not neighbors.

The circuit of Figure 10.5 creates a remotely entangled EPR pair in the following way. The Hadamard gate and first CNOT gate create an EPR pair between qubits 0 and 1, just as in the 2-qubit case (Equation 10.1). The second CNOT gate creates an EPR "triple" $\frac{1}{\sqrt{2}}(|000\rangle + |111\rangle)$ on the first three qubits. The third CNOT gate eliminates qubit 1 from the triple, leaving qubits 0 and 2 in an EPR pair, $\frac{1}{\sqrt{2}}(|000\rangle + |101\rangle)$. By induction, each subsequent pair of CNOT gates first creates an EPR triple among qubits 0, i, and $i+1$, and then removes qubit i from the triple, leaving qubits 0 and $i+1$ in an EPR pair. In this fashion, a remotely entangled EPR pair is eventually created among qubits 0 and $n-1$ via nearest-neighbor interactions.

In the absence of errors, the two computational basis states $|00\ldots0\rangle$ and $|10\ldots1\rangle$ occur upon measurement with a probability of $\frac{1}{2}$. All other states occur with probability 0. In the presence of errors, as discussed below, the probabilities of the two desired states will become less than $\frac{1}{2}$, and the probabilities of the other undesired states will become greater than 0.

Error Model

We first consider random and continuous gate errors. The physical basis for such errors is imprecision in the method used to apply gates to qubits. The implementation of gates for most known quantum computing technologies involves manipulation of electro-magnetic (EM) radiation pulses, and imprecision in the quantum control of these pulses manifests itself in under- or over-rotation of qubit states [59, 47, 23, 11, 13, 111, 50, 66, 90]. As a result, our error model has the general form of a 1-qubit unitary matrix,

$$U(\theta) = \begin{bmatrix} cos(\theta/2) & -sin(\theta/2) \\ sin(\theta/2) & cos(\theta/2) \end{bmatrix}$$

where θ is a rotation parameter that depends directly on the desired EM frequency [61]. Not shown are different phase factors in front of each of the four elements. These factors are easily set for the type of gate to be applied [61]. Modeling a faulty 1-qubit gate with continuous error only requires the addition of a small random ϵ error parameter to θ,

$$U_f(\theta, \epsilon) = \begin{bmatrix} cos(\theta/2 + \epsilon) & -sin(\theta/2 + \epsilon) \\ sin(\theta/2 + \epsilon) & cos(\theta/2 + \epsilon) \end{bmatrix}$$

where ϵ is normally distributed about 0 with standard deviation σ [34]. This model for continuous gate error was used to study the effectiveness of error correction codes in nearest-neighbor qubit arrays. It has been shown that for such error correcting codes to be most effective, ϵ must range between 10^{-5} and 10^{-7} [34]. In the nearest-neighbor chain setting for nuclear-spin technology, an ϵ of around 10^{-6} is considered reasonable [11, 13].

The 1-qubit continuous gate-error model can be extended to 2-qubit controlled gates as follows,

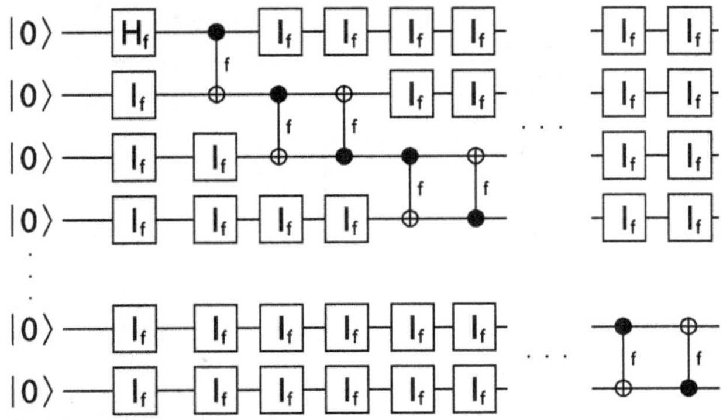

Fig. 10.6. Remote EPR-pair generation circuit of Figure 10.5 with gate and systematic errors added; a different randomly generated ϵ error parameter may be used for each gate.

$$V_f(\theta_0, \theta_1, \epsilon_0, \epsilon_1) = |0\rangle\langle 0| \otimes U_0(\theta_0, \epsilon_0) + |1\rangle\langle 1| \otimes U_1(\theta_1, \epsilon_1) \tag{10.2}$$

where U_0 is a faulty gate describing the action on the control qubit, and U_1 is a faulty gate describing the action on the target qubit. In the case of a faulty CNOT, U_0 is a faulty identity gate, and U_1 is a faulty X gate [61]. To reverse the order of the control and target qubits, the operands of the tensor products in Equation 10.2 are simply reversed. Modeling random continuous gate error in the remote EPR-pair generator (Figure 10.5) can be achieved by replacing the Hadamard gate with $H_f(\pi/4, \epsilon)$, and a CNOT gate with $CNOT_f(0, \pi/2, \epsilon_i, \epsilon_{i+1})$. Reversing the tensor products of $CNOT_f$ generates a faulty CNOT gate with reversed control and target qubits.

Besides gate errors, some quantum computing technologies are vulnerable to errors referred to as systematic errors or non-resonant effects [11, 13]. In the presence of systematic error, applying a gate to qubits i and $i+1$ has a small effect on all other qubits. To represent these small error effects when a faulty gate G_f is applied to one or more qubits, faulty identity gates of the form $I_i(0, \epsilon_i)$ are applied to all qubits not acted upon by G_f. This is a consistent model since in the error-free case, identity gates are implicitly present when a qubit is not acted upon by a gate.

The new form of the remote EPR-pair generation circuit, which includes gate and systematic errors, is shown in Figure 10.6. By inserting the faulty identity gates, the total number of gates in the faulty circuit is $(n-1)^2 + n$. Assuming the worst-case conditions for QuIDD-based simulation, a different randomly generated ϵ should be used in each faulty gate, including the systematic error identity gates, which reduces the number of repeated values that the QuIDDs compress. Other error models may cause the number of differ-

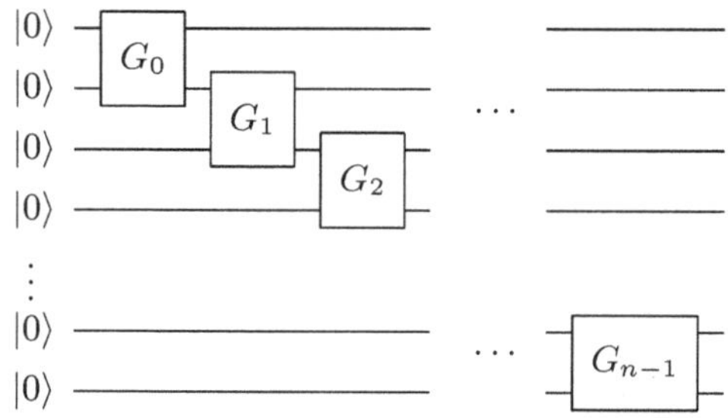

Fig. 10.7. Reduced version of the faulty remote EPR-pair generation circuit.

ent ϵ values to grow more rapidly, but such models should be no harder to simulate with QuIDDs than the case considered.

Jozsa describes a simple set of circuit reduction rules which can be applied to this circuit to analyze the difficulty of several different simulation techniques [43]. Essentially, all 1-qubit gates can be merged via matrix multiplication into neighboring 2-qubit gates, and such gates applied to the same qubits can also be merged. If this reduction is performed on the faulty circuit shown in Figure 10.6, the resulting circuit contains only 2-qubit gates which are applied in a cascade fashion as shown in Figure 10.7.

It is clear from the reduced circuit that after each 2-qubit gate is applied to qubits i and $i+1$, qubit i is no longer affects the computation and may be removed via the partial trace (with the exception of the first and last qubits). In fact, as the EPR pair propagates down to the last qubit, each intermediate qubit may be dynamically tensored in with the previous EPR pair to create the EPR triple. After applying the current 2-qubit gate, the middle qubit in the triple may be traced out. Using the dynamic tensor product and partial tracing technique discussed earlier, the space complexity of simulating this circuit is reduced to $O(1)$ and the time complexity is reduced to $O(n)$. Given that random, continuous errors normally cause QuIDDs to blow up exponentially in size, this optimization offers an asymptotic improvement in performance, as verified experimentally in the next subsection.

Next, we consider "collective dephasing" decoherence, which is known to be a major source of errors in ion-trap implementations[46]. This type of decoherence can be modeled as phase dampening and can be simulated with a single "environment qubit" which couples to each data qubit through a controlled-Y gate as shown in Figure 10.8 [61]. The angle parameter of the controlled-Y gate is a decoherence angle, where angles closer to π model a more rapid decoherence process [61]. For simplicity, our experiments assume the measurement outcome of the environment qubit is always in the "ground" state $|0\rangle$. From the perspective of simulation, the key fact to note is that, since the environment qubit is measured each time phase dampening is applied, it

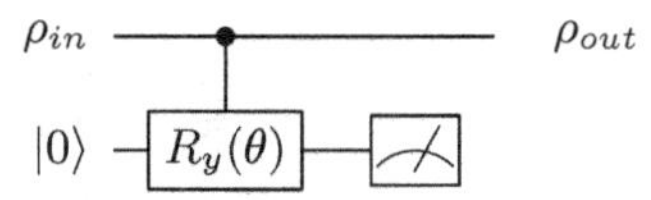

Fig. 10.8. Phase-dampening decoherence model involving an environment qubit.

assumes a classical state and is no longer coupled to the data qubit. Thus, decoherence only adds $O(1)$ runtime overhead using the dynamic tensoring and partial tracing technique for this circuit because the environment qubit can be removed via the partial trace.

It is important to note that p-blocked simulation, Vidal's slightly entangled technique, and tensor networks can all simulate this circuit efficiently in the presence of these errors as well. However, runtime overhead can be incurred due to swaps in Vidal's technique. These implications are discussed further in Chapter 11.

10.3 Empirical Validation

We used the dynamic tensoring and partial tracing technique in QuIDDPro to efficiently calculate the measurement outcome probabilities of all qubits in a faulty remote EPR-pair generation circuit. As noted in Section 10.2, a reasonable rotation error range for faulty gates is 10^{-5} to 10^{-7} [34, 11, 13]. Consequently, we consider three different cases in which random rotation errors are selected from normal distributions with ranges $\pm 10^{-5}$, $\pm 10^{-6}$ and $\pm 10^{-7}$, respectively. For each error distribution, we consider the remote EPR-pair generation circuit with gate errors only, and with gate and systematic errors together (decoherence is considered later). In each case, the error probability is calculated as $1 - P(|00\ldots0\rangle) - P(|10\ldots1\rangle)$ because, in the absence of errors, the probabilities of obtaining these outcomes should sum to 1. Also, since each gate is given its own randomly-generated rotation error, we compute the average of 100 different runs per error distribution.

Figures 10.9a-c depict the error probability due to gate errors only as a function of the number of qubits in the remote EPR-pair circuit. The data indicates that the probability of error increases linearly with the number of qubits. Figures 10.10a-c depict the probability of error due to gate and systematic errors as a function of the number of qubits. This data, however, indicates that in the presence of both gate and systematic errors, the error probability increases *quadratically* with the number of qubits. This asymptotic difference between the two cases is not surprising given that the number of faulty gates which must be simulated when modeling systematic errors is quadratic in the number of qubits.

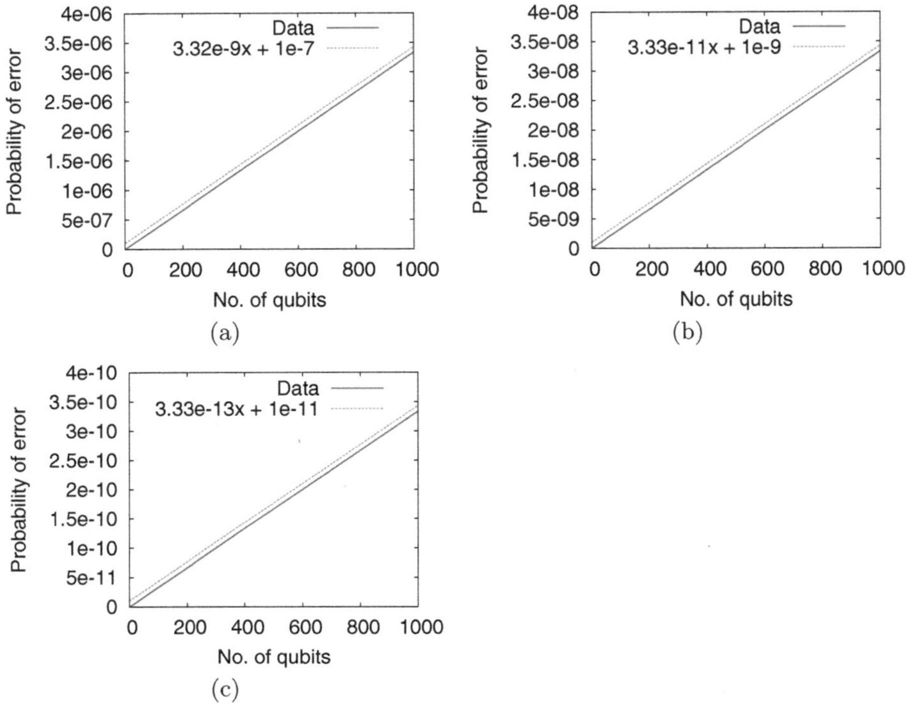

Fig. 10.9. Probability of error in the remote EPR-pair generation circuit due to gate errors only, as a function of the number of qubits. The rotation errors are randomly selected for each gate from normal distributions ranging from (a) $\pm 10^{-5}$, (b) $\pm 10^{-6}$, and (c) $\pm 10^{-7}$. The average of 100 runs is used for each distribution.

To model error growth as a function of gate number, the circuit size is fixed at 1000 qubits. This provides a good growth trend because the application of each pair of CNOT gates in sequence essentially models remote EPR-pair generation with one more qubit. In other words, applying CNOT gates up to and including qubits i and $i+1$ is equivalent to simulating a remote entanglement circuit with only $i+1$ qubits. Thus, the error trend for a 1000-qubit remote EPR-pair generation circuit as a function of the number of gates represents the trend for all remote EPR-pair generation circuits of size up to 1000 qubits. Figures 10.11a-c depict the probability of error in a 1000 qubit circuit in the presence of gate errors only, as a function of the number of gates. Note that the faulty identity gates are not counted, since systematic error is not an actual gate applied by the implementer of the quantum computer. The data indicates that the error probability increases linearly with the number of gates. Figures 10.12a-c show the error probability in the presence of gate and systematic errors. This data indicates that the number of errors increases quadratically with the number of gates. The similarity in growth trends is not

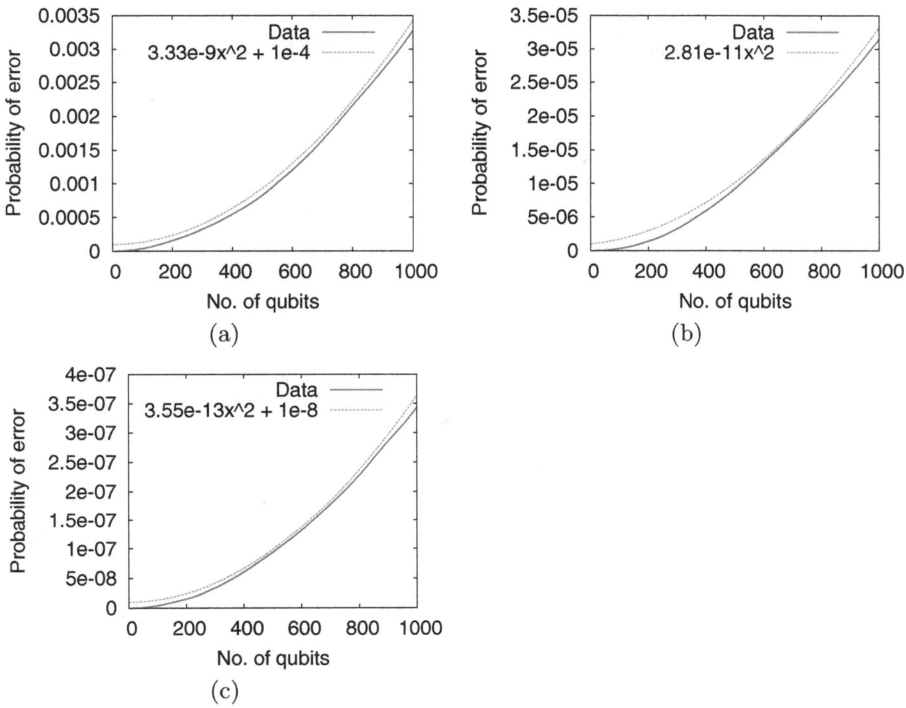

Fig. 10.10. Probability of error in the remote EPR-pair generation circuit, due to gate and systematic errors, as a function of the number of qubits. The rotation errors are randomly selected for each gate from normal distributions ranging from (a) $\pm 10^{-5}$, (b) $\pm 10^{-6}$ and (c) $\pm 10^{-7}$. The average of 100 runs is used for each distribution.

surprising, since the number of gates applied to each qubit is constant with respect to n.

In all cases, the error magnitude is very small, even though the error in the presence of systematic errors is several orders of magnitude larger than in the absence of such error. More importantly, the error growth as a function of the number of qubits and gates is sub-exponential. As a result, since remote EPR pair generation is a key step in quantum teleportation, error correction aimed at gate errors and/or systematic errors is probably unnecessary for teleporting a qubit state.

We also simulated the circuit in the presence of collective-dephasing decoherence error modeled as phase dampening. Whereas gate and systematic errors tend to increase the probability of measuring incorrect outcomes, phase dampening also skews the probability distribution of measuring the correct outcomes $|00\ldots0\rangle$ or $|10\ldots1\rangle$, i.e. the probabilities of measuring one correct state instead of the other are not equal. Thus, a better metric for these

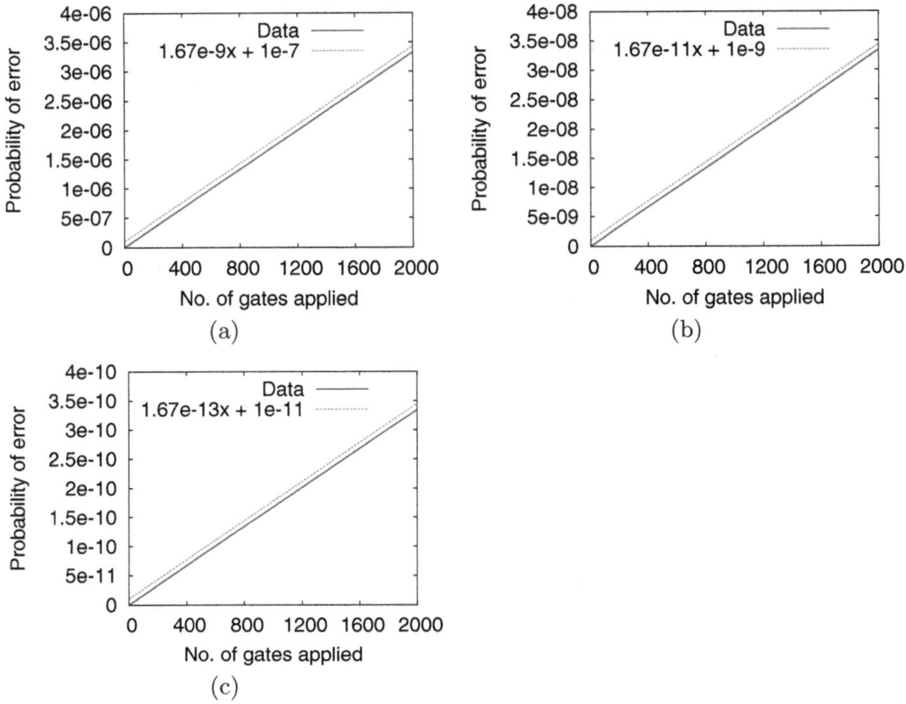

Fig. 10.11. Probability of error in the remote EPR-pair generation circuit due to gate errors only, as a function of the number of gates. The rotation errors are randomly selected for each gate from normal distributions ranging from (a) $\pm 10^{-5}$, (b) $\pm 10^{-6}$, and (c) $\pm 10^{-7}$. The average of 100 runs is used for each distribution.

experiments is the fidelity of the faulty state σ as compared to the correct state ρ. For density matrices, this is expressed as, $F(\rho, \sigma) = tr\sqrt{\rho^{1/2}\sigma\rho^{1/2}}$, where $F(\rho, \sigma)$ ranges between 0 (the states are completely different) and 1 (the states are equal) [61].

The first set of experiments simulate phase dampening alone with three different decoherence angles, 0.2, 1.2 and 3.0. The results are shown in Figures 10.13a-c, and they confirm that the fidelity drops much more quickly for decoherence angles closer to π. The second set of experiments simulate phase dampening with bang-bang error correction [109, 52, 108, 45, 68]. There are many ways to define the bang-bang corrective operators. In these experiments, the "universal decoupling" sequence is used, which alternates between the Pauli X and Z operators after every gate is applied [108, 45]. Compared to corrective operators that involve negations of the decoherence operator itself [109], this choice is arguably more realistic and useful to experimental physicists since it requires no knowledge of the Hamiltonian representing the underlying decoherence process. As shown in Figures 10.14a-b, this set

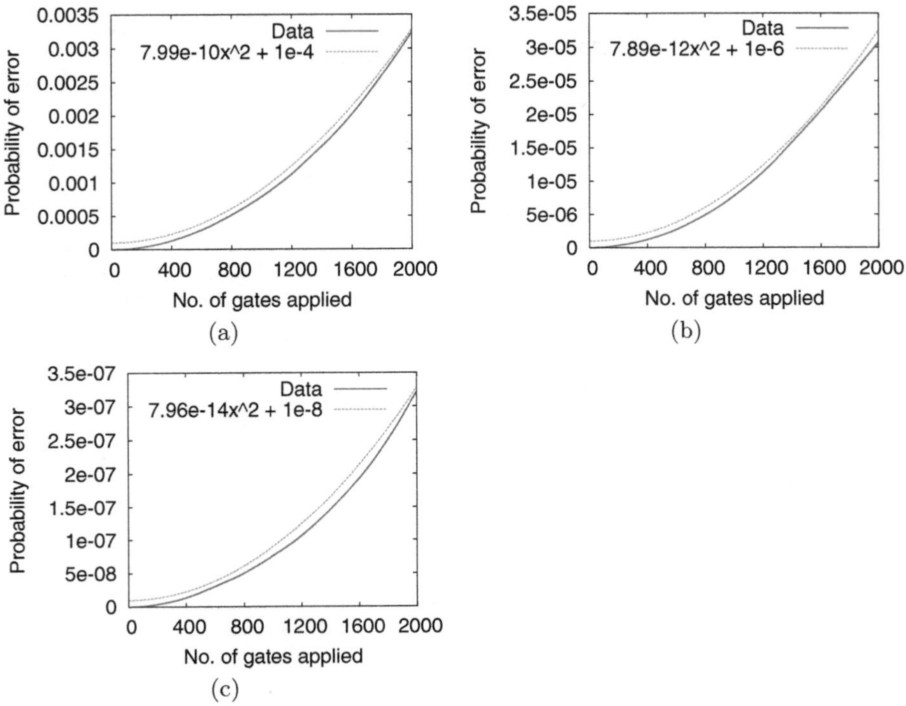

Fig. 10.12. Probability of error in the remote EPR-pair generation circuit due to gate error and systematic error, as a function of the number of gates. The rotation errors are randomly selected for each gate from normal distributions ranging from (a) $\pm 10^{-5}$, (b) $\pm 10^{-6}$, and (c) $\pm 10^{-7}$. The average of 100 runs is used for each distribution.

of bang-bang operators is extremely effective for this particular circuit. Unlike the previous results, the fidelity never reaches 0. However, Figure 10.14c shows that the extremely rapid decoherence process modeled by decoherence angle 3.0 is not effectively dealt with by this choice of operators. Since this angle is so close to π, there may be no practical way to cope with such a rapid decoherence process using bang-bang operations.

An open question is how effective bang-bang operators are in the presence of gate error due to imprecision [110, 45]. Figures 10.15a-c provide data from the last set of experiments which model phase dampening, the bang-bang operators used in the previous experiments, and a gate error range of $\pm 10^{-5}$ (the worst-case range). Interestingly enough, the bang-bang operators are indeed able to cope with decoherence angles 0.2 and 1.2 as before, suggesting that gate imprecision may not be a significant problem for bang-bang error correction.

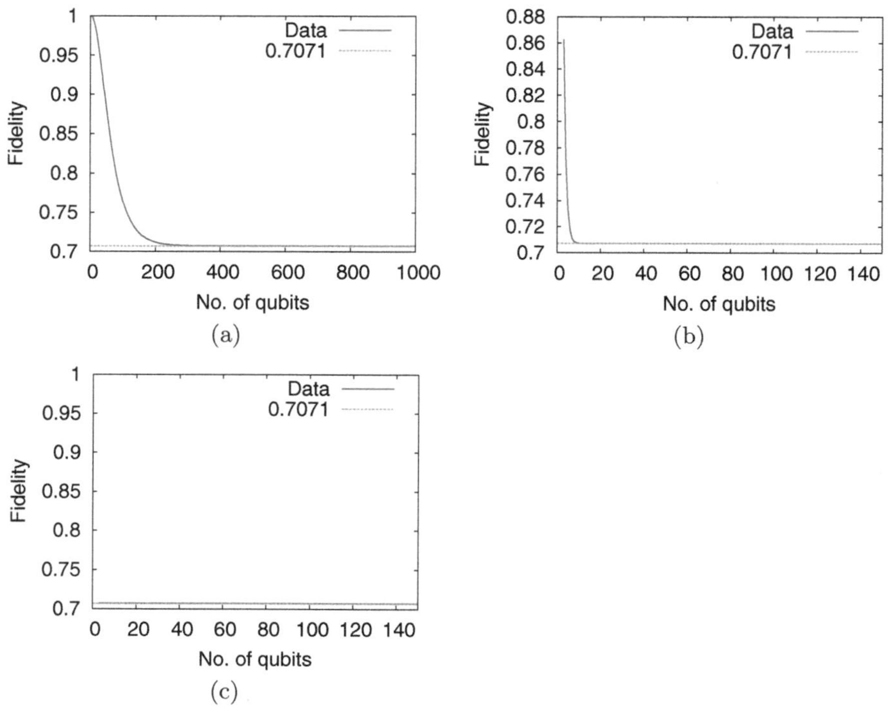

Fig. 10.13. State fidelity in the remote EPR-pair generation circuit with decoherence as a function of the number of qubits. The decoherence angles used are (a) 0.2, (b) 1.2, and (c) 3.0. Results are only shown for up to 140 qubits for (b) and (c) since the fidelity drops to approximately $1/\sqrt{2}$ quickly.

10.4 Summary

We have discussed two important language-level techniques for state-vector and density-matrix simulation. In the case of state-vector simulation, the QuIDDPro simulator automatically detects when to apply specialized gate-specific algorithms depending on the expressions used. With such optimizations, QuIDDs are comparable with stabilizer simulation for several large benchmarks, especially when non-stabilizer gates (e.g., Toffoli) are introduced. In the case of density-matrix simulation, some language features are used to introduce qubits only when needed, and remove them when they no longer affect the qubits of interest. This approach can also characterize gate, systematic, and decoherence errors, as well as bang-bang error correction in the remote EPR-pair generation circuit.

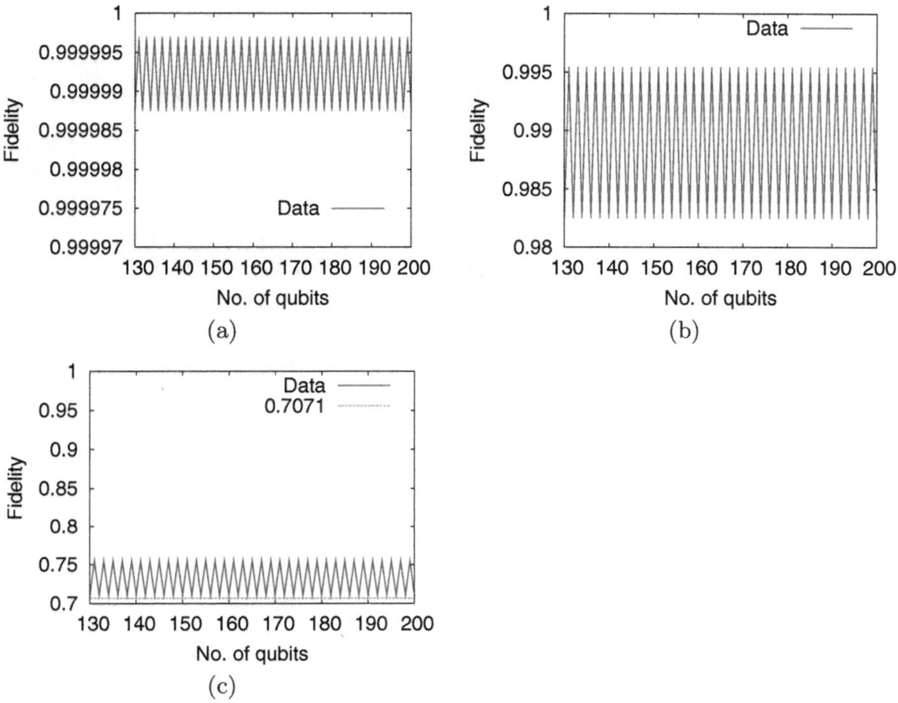

Fig. 10.14. State fidelity in the remote EPR-pair generation circuit with decoherence as a function of the number of qubits. Bang-bang pulses from the universal decoupling sequence are used to correct the state after every gate is applied. The decoherence angles used are (a) 0.2, (b) 1.2, and (c) 3.0. Results are given from 130 to 200 qubits so that the periodic nature of the data is easily viewed. The trends continue through 1000 qubits.

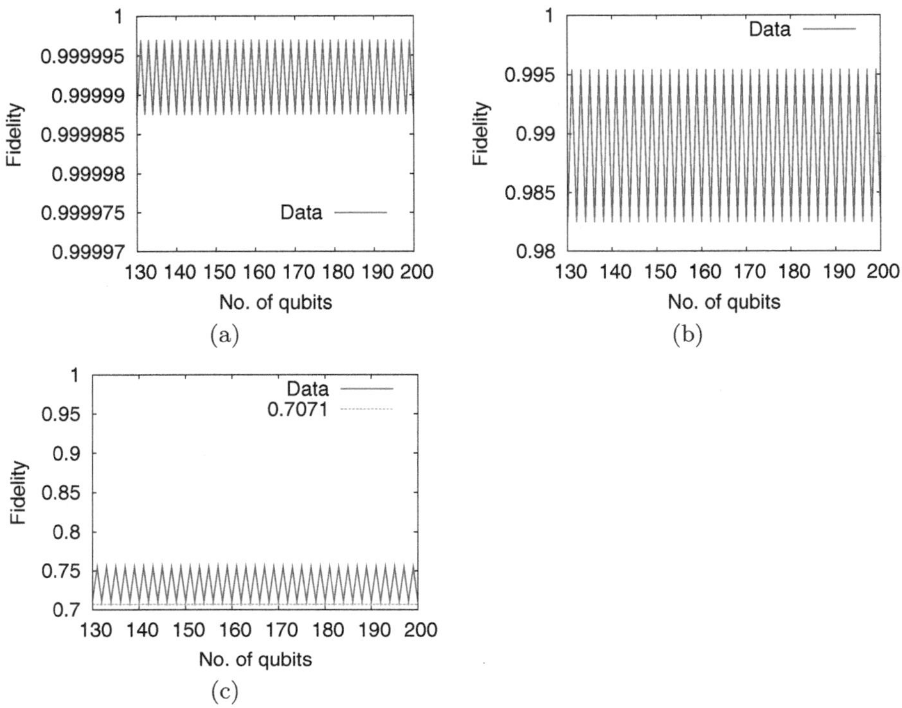

Fig. 10.15. State fidelity in the remote EPR-pair generation circuit with decoherence as a function of the number of qubits. Faulty bang-bang pulses from the universal decoupling sequence with an error range $\pm 10^{-5}$ are used to correct the state after every gate is applied. The decoherence angles used are (a) 0.2, (b) 1.2, and (c) 3.0. Results are given from 130 to 200 qubits so that the periodic nature of the data is easily viewed. The trends continue through 1000 qubits.

11

Closing Remarks

In general, quantum circuit simulation requires runtime and memory resources that grow exponentially with the number of qubits simulated. Some circuits can be simulated quickly, but are unlikely to exhibit quantum speed-ups. To this end, quantum circuit simulation often suggests certain limitations of quantum circuits and algorithms.

Quantum simulation can also aid in quantum engineering efforts by evaluating and sanity-checking small versions of quantum circuits before actual implementation. Quantum circuits are significantly more complicated than classical digital logic circuits, and their properties are difficult to capture with traditional CAD techniques. One of the most important of these properties is the fragile nature of quantum information. Quantum states are often damaged over time by several types of gate-specific and environmental errors, which experimental physicists find difficult to characterize. Additionally, the notion of equivalence, while trivial in the classical case, takes on a surprisingly rich set of interpretations in the quantum case, offering several computational challenges of varying complexity. As a result, useful quantum CAD tools must incorporate special models and efficient, classical simulation techniques to overcome these obstacles for classes of circuits with practical value.

The techniques discussed in this book efficiently simulate various classes of quantum circuits, but how much these classes overlap in practice remains unclear. For example, the remote EPR pair generation circuit analyzed in Chapter 10 can also be simulated by Vidal's technique and by tensor networks. Decoherence errors can induce $\Omega(n)$ wire-swaps when using Vidal's method, but this overhead appears avoidable by dynamically concatenating single-qubit tensors to Vidal's tensor decomposition, similar to what is done with QuIDDs and dynamic tensor products in Chapter 10. The concatenation should be straightforward when introducing a new qubit in the ground state since there is no entanglement between the current state and the new qubit ($\chi = 1$). The partial trace would also be required to remove the environment qubit after decoherence and measurement are applied.

This example illustrates several key points about practical quantum circuit simulation. First, regardless of the back-end simulation technique, the front-end language and supporting functionality are important. Without the power to express certain simulation short-cuts and specialized gates, a great deal of computational resources may be wasted. In some cases, such unnecessary runtime and memory overhead can grow asymptotically.

Second, characterizations of quantum circuits that can be simulated efficiently would be very useful, as well as descriptions of overlaps between classes of such circuits. This can lead to new theoretical results, help improving algorithms and optimize simulation software. For instance, recent research combines some of the simulation techniques discussed in this book. Shi, Duan and Vidal replace the tensors in Vidal's decomposition with tensor networks [81]. Whereas Vidal's technique alone efficiently simulates one-dimensional quantum many-body systems [107] and tensor contractions alone efficiently simulate tensor networks with low treewidth, the hybrid approach efficiently simulates quantum many-body systems of arbitrary dimension so long as their tensor-network representation has low treewidth [81].

In addition to enabling hybrid approaches, a better understanding of overlaps among known techniques could elucidate relevant data structures. For example, Anders and Briegel's replacement of (dense) bit tables with a (sparse) graph-based data structure in the stabilizer formalism reduces the complexity of simulating that particular class of quantum circuits [5]. QuIDDs too can be viewed as a replacement of explicitly stored matrices and vectors. Such data structures draw from many seemingly disjoint areas of computer science and engineering, ranging from theoretical algorithmic analysis to heuristic CAD for classical digital logic design. It is likely that quantum circuit simulation can benefit from additional data structures, algorithms, and heuristics.

Lastly, identifying representative benchmark circuits with various types of physically realistic errors and simulating their performance is crucial to expanding the practical value of existing simulation techniques. In addition to theoretical aspects of quantum simulation, experimental physicists, like electrical engineers who design modern CPUs, have very practical requirements for specific applications. In developing quantum circuit CAD tools, simulation algorithms and asymptotic complexity results should be treated as a foundation on which to build robust, efficient and versatile software packages. QuIDDPro is an end-to-end project which started with algorithmic contributions to simulation and culminated in a rich software package aimed at providing physicists and other researchers in the field with a useful tool.

Research covered in this book lays foundations for the development of quantum circuit simulation techniques. We hope that the QuIDD data structure and the QuIDDPro software package illustrate how to pursue such development. The continued efforts of many researchers in the emerging field of quantum information processing, combined with decades of experience accumulated in classical circuit CAD methods, will lead to even more powerful tools for analyzing and synthesizing quantum circuits.

Appendix A

QuIDDPro Simulator

QuIDDPro is a quantum circuit simulator developed around the QuIDD data structure and QuIDD-based algorithms. It provides numerous built-in functions and language features which make QuIDDs transparent and easy to use. This appendix provides a brief overview of how to run the simulator, as well as a language reference.

A.1 Running the Simulator

QuIDDPro can be run in two modes, batch and interactive. In batch mode, the user supplies the simulator with an ASCII text file containing the script code to be executed. The text file can be provided as an argument in the command line to the simulator executable or redirected to standard input, as in the following examples:
File "my_code.qpro" passed as an argument:

```
% ./qp my_code.qpro
```

File "my_code.qpro" redirected to standard input:

```
% ./qp < my_code.qpro
```

Note that although the examples use a ".qpro" extension in the filenames, any valid filename will do.

Interactive mode is triggered when the simulator executable is given no arguments at the command line. In this mode, the simulator will be started and produce a prompt to await input from the user as follows:

```
% ./qp
QuIDDPro>
```

Similar to MATLAB, valid lines of code may be typed at the prompt and executed when the return or enter key is pressed, i.e., when a newline character is given as input. The command "quit" is issued to exit the simulator. Multiple expressions may be placed in a single line separated by semicolons. An example of this method of input is as follows:

```
QuIDDPro> a = pi/3; r_op = [cos(a/2) -i*sin(a/2); -i*sin(a/2) cos(a/2)]
```

```
r_op =
 0.866025 0-0.5i
 0-0.5i 0.866025
```

In this example, a 1-qubit rotational X operator matrix is created with the θ parameter $\pi/3$. Notice that only the value of the variable "r_op" is printed out. In general, the value of the last expression is printed out for an input line containing multiple expressions separated by semicolons. However, the other expressions are still computed. In this example, for instance, the variable "a" will contain the value $pi/3$, even though this result is not printed out. This is true because the definition of "r_op" depends on the value of "a." In addition to providing the means to place multiple expressions on the same line, semicolons can be used more generally to suppress output to the screen. If screen output for any particular expression is not desired, simply place a semicolon at the end of the expression to compute it silently. MATLAB behaves in the same fashion.

QuIDDPro contains a number of built-in functions and predefined variables. A listing of such functions and variables can be found in Section A.3. Notice that in the last example, the predefined variables "pi" and "i" are used. "pi" contains the value π (to a large number of digits), while "i" contains the value $0 + i$. Predefined variables can be overwritten by the user. In addition to the predefined variables just mentioned, the built-in functions "cos" and "sin" were also used in the last example.

To demonstrate the use of built-in functions further, consider the example:

```
QuIDDPro> r_op = rx(pi/3, 1)
r_op =
 0.866025 0-0.5i
 0-0.5i 0.866025
```

Here, the built-in function "rx" is used to create the same matrix that was created in the previous example, namely the 1-qubit rotational X operator. QuIDDPro provides a number of such functions to create commonly used operators. See Section A.3 for more details.

Although interactive mode is useful for quick calculations, it may not be preferable for non-trivial pieces of code that are reused many times. Thus, batch mode is highly recommended for most contexts. In the next example, we demonstrate how to use QuIDDPro to simulate a quantum circuit in batch mode. The code shown here can be placed into a file for execution at any time. Consider the canonical decomposition of a two-qubit unitary operator U described in [24]. U can be expressed as:

$$U = (A_1 \otimes B_1)e^{i(\theta_x X\otimes X+\theta_y Y\otimes Y+\theta_z Z\otimes Z)}(A_2 \otimes B_2)$$

subject to the constraint that $\frac{\pi}{4} \geq \theta_x \geq \theta_y \geq |\theta_z|$ and A_1, A_2, B_1, and B_2 are one-qubit unitary operators.

Suppose we wish to simulate a quantum circuit in which some 2-qubit unitary operator U is to be applied to two qubits in the density-matrix state $|10\rangle\langle 10|$. Further, suppose that U must be computed given the canonical decomposition parameters $\theta_x = 0.702$, $\theta_y = 0.54$, and $\theta_z = 0.2346$. Additionally, we are given that A_1 is a one-qubit Hadamard operator, A_2 is X, B_1 is I, and B_2 is Y. This can be implemented with the following code:

```
theta_x = 0.702;
theta_y = 0.54;
theta_z = 0.2346;
A1 = hadamard(1);
A2 = sigma_x(1);
B1 = identity(1);
B2 = sigma y(1);
```

Next, U can be computed with the code:

```
Xpart = theta_x*kron(sigma_x(1), sigma_x(1));
Ypart = theta_y*kron(sigma_y(1), sigma_y(1));
Zpart = theta_z*kron(sigma_z(1), sigma_z(1));
U = kron(A1, B1)*expm(i*(Xpart + Ypart + Zpart))*kron(A2, B2)
```

U is then applied to the density matrix state $|10\rangle\langle 10|$ with the code:

```
state = cb(''10'');
final state = U*(state*state')*U'
```

Deterministic measurement can be performed to eliminate the correlations associated with each qubit:

```
q_index = 1;
while (q_index < 3)
final_state = measure(q_index, final_state);
q_index = q_index + 1;
end
measured_state = final_state
```

U can also be applied very easily to the state-vector representation of the state if it is preferred to the density-matrix representation. In addition, the probability of measuring a 1 or 0 for any qubit in the state vector can be computed using other measurement functions:

```
final_state_v = U*state
p0_qubit1 = measure_sv0(1, final_state_v)
p1_qubit1 = measure_sv1(1, final_state_v)
p0_qubit2 = measure_sv0(2, final_state_v)
p1 qubit2 = measure sv1(2, final state v)
```

Probabilistic measurement can also be performed on both density matrices and state vectors. See `pmeasure` and `pmeasure_sv` in Section A.3 for more details.

Upon execution of the above script, the output is:

```
U =
-0.110927-0.0265116i -0.0530448-0.222078i -0.650863+0.15556i 0.162218-0.678733i
-0.162218+0.678733i  0.650863-0.15556i   0.0530448+0.222078i 0.110927+0.0265116i
-0.110927-0.0265116i 0.0530448+0.222078i 0.650863-0.15556i   0.162218-0.678733i
0.162218-0.678733i   0.650863-0.15556i   0.0530448+0.222078i -0.110927-0.0265116i
final_state =
0.447822   2.15483e-05+0.152794i -0.447822    2.15483e-05+0.152794i
```

```
2.15483e-05-0.152794i 0.0521324  -2.15483e-05+0.152794i   0.0521324
-0.447822 -2.15483e-05-0.152794i  0.447822   -2.15483e-05-0.152794i
2.15483e-05-0.152794i 0.0521324  -2.15483e-05+0.152794i   0.0521324
measured_state =

 0.447822     0          0         0
    0     0.0521324     0          0
    0         0      0.447822      0
    0         0          0     0.0521324

final_state_v =
-0.650863+0.15556i
0.0530448+0.222078i
0.650863-0.15556i
0.0530448+0.222078i

p0_qubit1 =
0.499955
p1_qubit1 =
0.499955
p0_qubit2 =
0.895644
p1_qubit2 =
0.104265
```

Although the examples in this section demonstrate scripts that use small numbers of qubits, the real power of QuIDDPro lies in simulating quantum-mechanical systems with many quantum states (10 or more). The benchmark circuit large_h.qpro, for instance, applies a 50-qubit Hadamard operator to a density matrix of 50 qubits. steaneX.qpro and steaneZ.qpro demonstrate error correction in quantum circuits of 12 and 13 qubits, respectively. On a single-processor workstation, each of these scripts requires less than 5 seconds to run and less than 0.5 MB of peak memory usage.

A.2 Functions and Code in Multiple Files

QuIDDPro supports user-defined functions via the "m-file" model commonly used in MATLAB. Specifically, a function call to a user-defined function may appear anywhere as long as the function body is contained in a separate file in the working directory. The name of the file containing the function body must be the same as the function name with ".qpro" or ".qp" appended. To illustrate, consider the following script which uses an oracle function to implement a simple instance of Grover's algorithm shown on page 256 of [61]. Notice that Dirac-style syntax maybe used for state-vector QuIDDs.
(*examples/functions/simple_grover.qpro*)

```
|state:> = cb(''001'');
|state:> = hadamard(3)*|state:>;
```

```
|state:> = oracle(|state:>);
|state:> = cu_gate(hadamard(1), ''xxi'')*|state:>;
|state:> = cu_gate(sigma_x(1), ''xxi'')*|state:>;
|state:> = cu_gate(hadamard(1), ''ixi'')*|state:>;
|state:> = cu_gate(sigma_x(1), ''cxi'')*|state:>;
|state:> = cu_gate(hadamard(1), ''ixi'')*|state:>;
|state:> = cu_gate(sigma_x(1), ''xxi'')*|state:>;
|state:> = hadamard(3)*|state:>
```

(*examples/functions/oracle.qpro*)

```
function |new_state:> = oracle(curr_state)
   |new state:> =_cu gate(sigma x(1), ''ccx'')*|curr state:>;
```

The user-defined function is "oracle" with its function body defined in the file "oracle.qpro." The other functions used are part of the QuIDDPro language (see Section A.3 for more details). Notice that in this particular example, the QuIDD "state" is passed as a function argument. In QuIDDPro, a QuIDD function argument only requires $O(1)$ memory space, because a pointer to the head of the QuIDD is passed to a function rather than the entire QuIDD. The same holds true for returning QuIDDs from a function. Thus, passing QuIDD arguments and return values is extremely efficient. In general, a user-defined function can contain any number of parameters which can be any combination of QuIDDs or complex numbers. Arguments passed as parameters to functions are not modified by the function (i.e. pass-by-value is always used).

Unlike MATLAB, QuIDDPro functions must have only one return variable (a function that returns nothing is also not allowed). If the function is intended to return no values, such as a diagnostic printing function, then a dummy variable can be used for the return variable. The return variable need not be used in the function body, and when this occurs, it is automatically assigned a value of 0. A semicolon can be appended to the function call to suppress the output of the 0 value. When multiple return values are desired, they can be stored together in a matrix. Thus, requiring a single return variable does not actually restrict the number of values that can be returned.

Like MATLAB and other languages, variables declared locally in a function body exist in their own scope. In other words, variables declared in a function body are undefined upon leaving the function body. By the same token, such variables do not overwrite the values of variables with the same name declared outside the function body.

In addition to functions, QuIDDPro supports the *run* command. Like its MATLAB counterpart, this command runs script code contained in another file. In the following example, the same circuit as before is simulated, but this time the run command is used instead of a user-defined function.

(*examples/run/simple_grover.qpro*)

```
run ''oracle_def.qpro''
state = cb(''001'');
state = hadamard(3)*state;
```

```
state = oracle*state;
state = cu_gate(hadamard(1), ''xxi'')*state;
state = cu_gate(sigma_x(1), ''xxi'')*state;
state = cu_gate(hadamard(1), ''ixi'')*state;
state = cu_gate(sigma_x(1), ''cxi'')*state;
state = cu_gate(hadamard(1), ''ixi'')*state;
state = cu_gate(sigma_x(1), ''xxi'')*state;
state = hadamard(3)*state
```

(*examples/run/oracle_def.qpro*)

```
oracle = cu_gate(sigma_x(1), ''ccx'');
```

Notice that the run command does not introduce a new scope. All variables declared in a run file exist in the current scope. As a result, the run command is ideal for declaring variables which can be re-used in multiple projects. Also, there is no constraint on where a run command may appear other than that it may not be placed within an explicit matrix.

A.3 Language Reference

This section provides a reference for the QuIDDPro input language. Although the language is similar to that of MATLAB, there are many functions in QuIDDPro specific to quantum mechanics which do not exist in MATLAB. There are also many functions in MATLAB which are not supported by QuIDDPro. Additionally, some of the functions that have the same names as those in MATLAB have slightly different functionality from their MATLAB counterparts. New language features will be added in future versions of the QuIDDPro simulator, and we welcome user suggestions.

<table>
<tr><td>==</td><td>~=, !=</td><td><</td><td><=</td></tr>
<tr><td>></td><td>>=</td><td>&&</td><td>||</td></tr>
<tr><td>+</td><td>-</td><td>*</td><td>/</td></tr>
<tr><td>=</td><td>'</td><td>(...)</td><td>^</td></tr>
</table>

Operations

cutoff_val	i
output_prec	pi
qp_epsilon	r2
r3	

Predefined variables

[...]	;	a(n, k)
a($n_1, n_2, n_3, \ldots$)	else	elseif
function	if	run
tic	toc	while
for	end	

Language features

atan	cb	cnot	conj
cos	cps	cu_gate	dump_dot
echo	exp	expm	eye
fredkin	gen_amp_damp	hadamard	identity
kron	norm	measure	measure_sv
measure_sv0	measure_sv1	pmeasure	pmeasure_norm_sv
pmeasure_sv	proj0	proj1	projplus
ptrace	px, Px	py, Py	pz, Pz
quidd_info	rand	round	rx, Rx
ry, Ry	rz, Rz	sigma_x	sigma_y
sigma_z	sin	sqrt	swap
toffoli	zeros		

Built-in Functions

- [...] defines a matrix explicitly. Expressions are placed between the brackets. Elements in the same row are separated by whitespace (including newlines) or commas, while rows are separated by one or more semicolons. The brackets can be nested within other brackets (matrices within matrices).
- # starts a one-line comment. Everything from the # symbol to the first newline is ignored. An alternative comment symbol is %.
- % starts a one-line comment. Everything from the % symbol to the first newline is ignored. An alternative comment symbol is #.
- ′ returns the complex-conjugate transpose of a matrix. For example, $[1\ 2;\ 3+2i\ 4]' \rightarrow [1\ 3-2i;\ 2\ 4]$
- == equality operation that returns 1 if the two expressions compared are equal; otherwise it returns 0. Comparison between matrices is supported. A complex number and a matrix are considered not equal unless the matrix has dimensions 1×1 and contains a number equal to the one being compared to.
- ~= inequality operation that performs the complement function of ==.
- != an alternative symbol for ~=.
- < less than operation. It returns 1 if the left-hand expression is less than the right-hand express; otherwise it returns 0. It can only be used to compare numbers. For numbers with nonzero imaginary components, only the real parts are compared.
- <= less than or equal operation. It returns 1 if the left-hand expression is less than or equal to the right-hand express; otherwise it returns 0. It can only be used to compare numbers. For numbers with nonzero imaginary components, only the real parts are compared.
- > greater than operation. It returns 1 if the left-hand expression is greater than the right-hand express; otherwise it returns 0. It can only be used to compare numbers. For numbers with nonzero imaginary components, only the real parts are compared.
- >= greater than or equal operation. It returns 1 if the left-hand expression is greater than or equal to the right-hand express; otherwise it

returns 0. It can only be used to compare numbers. For numbers with nonzero imaginary components, only the real parts are compared.

- && logical AND connective. It returns 1 if both sides of the operator evaluate to 1; otherwise it returns 0. It can only be used to compare numbers with nonzero imaginary components.
- || logical OR connective. It returns 1 if either side of the operator evaluates to 1; otherwise it returns 0. It can only be used to compare numbers with nonzero imaginary components.
- $+$ addition operation. For complex numbers, it returns the sum of the numbers. For matrices, it returns the element-wise addition of both matrices (both matrices must have the same number of rows and columns). When a matrix is added to a complex number, the complex number is added to each element of the matrix as a scalar.
- $-$ subtraction operation. For complex numbers, it returns the difference of the numbers. For matrices, it returns the element-wise difference of both matrices (both matrices must have the same number of rows and columns). When a matrix is subtracted from a complex number or vice-versa, scalar subtraction is performed element-by-element. When there is no left-hand expression, it is treated as a unary minus applied to the right-hand side expression. Within a matrix definition, for example $[1\ \ -2]$, the minus sign is treated as a unary minus. However, in $[1{-}2]$ and $[1\ -\ 2]$, the minus sign is treated as the binary minus expression. Parenthesis can be used to force the minus sign to be treated one way or the other.
- $*$ multiplication operation. For complex numbers, it returns the product of the numbers. For matrices, matrix multiplication is performed (as opposed to element-wise multiplication). Scalar multiplication is performed when a matrix and a complex number are multiplied together.
- / division operation. For complex numbers, it returns the quotient. Unlike the C language, integer division is *not* performed if the operands are both integer values. Double-precision floating-point division is always performed. For matrices, element-wise division is performed (both matrices must have the same number of rows and columns). When a matrix is divided by a complex number, scalar division is performed. However, a complex number may not be divided by a matrix.
- $=$ assignment operation. It assigns the value of an expression (right-hand side) to a variable (left-hand side). The expression may result in either a complex number or a matrix. The left-hand side expression must be a variable name (it must start with a letter and contain only alphanumeric characters and optionally underscores). Variables can be assigned "on-the-fly." In other words, unlike languages like C/C++, variables are not declared nor typed in any way prior to their first assignment. However, a variable must be assigned a value before it can be used in an expression. Similar to languages such as C/C++, an assignment expression returns a value just like any other expression, namely the value that was assigned to the variable on the left-hand side. Therefore, statements such as $x =$

$y = 3 + 4i$ are valid. In statements like these, if output is not suppressed, the value of the leftmost variable will be output to the screen. Although the other variables assigned values will not be output to the screen, they are still assigned their values. Another important note is that even though string literals appear as arguments in some functions, including *cu_gate* and *echo*, assignment of a string literal to a variable is not yet supported.

- $\wedge$ exponentiation operation for complex numbers. It returns the expression on the left-hand side of the $\wedge$ raised to the power of the expression on the right-hand side. For matrix exponentiation, see the `expm` function.
- (...) forces precedence for an expression as in any other programming language. An expression within the parentheses is evaluated before evaluating expressions outside of the parentheses.
- ; the semicolon suppresses output of an expression. For example, $x = 1$ stores the value of 1 in the variable x and outputs $x = 1$ to standard output, whereas $x = 1$; also stores the value of 1 in the variable x but would not output anything to standard output. When a semicolon appears in a matrix definition, it has a different meaning entirely. Within a matrix definition, a semicolon denotes the end of a row.
- `a(`*n, k*`)` if `a` is a variable containing a matrix, then this expression returns the element indexed by the row index n and the column index k. Numbering of indices starts at 1. Unlike languages such as MATLAB, this expression may not be used to assign values to elements of a matrix. It may only be used to read a particular element from a matrix (e.g. `x = a(`*1, 2*`) + 2` is valid, but `a(`*1, 2*`) = 3+2` is not). Future versions may support this, however, if there is demand for such functionality. n and k must be complex numbers with no imaginary components. n and k must also each be within $10E - 5$ of an integer value (e.g. 9.99999, 1.00001, and 3 are valid, but 4.5 is not), and values that are within this threshold are rounded to the nearest integer (e.g. 9.99999 is interpreted as 10). In addition, n and k must each be at least 1 after rounding.
- `a(`$n_1, n_2, n_3, \ldots$`)` if `a` is not a variable containing a matrix, it is considered to be user-defined function call. n_1, n_2, and n_3 are function arguments that can be expressions or variables of any type. There is no constraint on the number of arguments. Also note that passing QuIDD arguments and QuIDD return values only requires $O(1)$ memory since only a single pointer to the head of a QuIDD needs to be passed. Arguments passed as parameters to functions are not modified by the function (i.e. pass-by-value is always used). See Section A.2 for more details.
- `atan(`n`)` returns the arc tangent of the expression n passed as an argument. If n is a matrix, it returns a matrix containing the element-wise arc tangent of n.
- `cb(`"..."`)` returns a computational basis state vector. The string literal argument consists of a sequence of any number and combination of '0' and '1' characters. The string is parsed from left to right. Each '0' causes a $|0\rangle$ to be tensored into the vector, and each '1' causes a $|1\rangle$ to be tensored into

the vector. `cb` can easily be used to create density matrices by using it in conjunction with the complex-conjugate transpose operation ('), matrix multiplication, and scalar operations.

- `cnot("...")` returns a 2-qubit controlled-NOT (CNOT) gate matrix. This is a faster, specialized version of `cu_gate`. If a controlled gate matrix with different numbers of controls/targets and/or a different action (U operator) is desired, then use the more general `cu_gate` function. The argument of `cnot` is a string literal using the same gate specification syntax as `cu_gate`. However, the only valid parameters accepted by `cnot` are *'cx'* and *'xc'*, since these string specifications are the only possible strings that produce a valid 2-qubit CNOT gate matrix. For example, `cnot('cx')` produces a CNOT gate matrix with the control on the "top" wire and the action (X operator) on the "bottom" wire. For a discussion of how the concept of wires relates to creating controlled gate matrices, see `cu_gate`.
- `conj(`*n*`)` returns the complex-conjugate of the expression n passed as an argument. n can be a complex number or a matrix.
- `cos(`*n*`)` returns the cosine of the expression n passed as an argument. If n is a matrix, it returns a matrix containing the element-wise cosine of n.
- `cps(`*n*`)` returns an n-qubit phase-shift (also called conditional phase shift or CPS) gate matrix. n must be a complex number with no imaginary component. n must also be within $10E-5$ of an integer value (e.g. 9.99999, 1.00001, and 3 are valid, but 4.5 is not), and values that are within this threshold are rounded to the nearest integer (e.g. 9.99999 is interpreted as 10). In addition, n must be at least 1 after rounding. Always use this function instead of explicitly defining your own CPS matrix. This function is asymptotically faster and uses asymptotically less memory than defining the matrix explicitly. The phase-shift gate is particularly useful in Grover's quantum search algorithm [38].
- `cu_gate(`*a*`, "...")` is a generalized controlled-U gate matrix creation function. It returns a controlled or uncontrolled gate matrix given an action matrix (a) and a string literal with the gate specification (the second argument contained in "s). The string literal consists of a sequence of characters. The idea is for the string literal to specify what the gate should do to each "wire" in a quantum circuit. When conceptualizing a quantum circuit graphically and reading top-down, the first character corresponds to the first qubit wire, the second character corresponds to the second qubit wire, etc. Each character can take one of four possible values. 'i' denotes the identity, which means that the gate does nothing to the wire at that location. 'x' denotes an action, which means that the matrix specified by the argument a is applied to the wire at that location. 'c' denotes a control, which means that the wire at that location is used as a control on any 'x' wire (a $|1\rangle$ state forces a to operate on any 'x' wire, whereas a $|0\rangle$ causes nothing to happen on any 'x' wire). 'n' is a negated control, which is the opposite of 'c' (a $|0\rangle$ state forces a to operate on any 'x' wire, whereas a $|1\rangle$ causes nothing to happen on any 'x' wire). Any sequence of these

characters may be used. Although there is no "actual" circuit, the string characters allow a user to conceptualize a circuit and construct a matrix which operates on the wires in that conceptualized circuit. a may be a matrix that operates on more than one qubit as long as one or more blocks of contiguous 'x' characters appear such that the size of each block is equal to the number of qubits operated on by a. For examples, see steaneX.qpro and steaneZ.qpro under the examples/nist/ subdirectory. Always use this function instead of defining your own gates explicitly, since it is asymptotically faster and uses asymptotically less memory. Since `cu_gate` must parse the input specification string, other functions such as `hadamard` and `cps` should be used instead of `cu_gate` for specific gates because they do not perform any parsing and are therefore a bit more efficient. An alternative function name for `cu_gate` is `lambda`. Also see the alternative, condensed version of `cu_gate` discussed next. The alternative version may be preferable for circuits with many qubits.

- `cu_gate(`*a, "...", n*`)` is an alternative syntax for `cu_gate` which takes a condensed string literal "...". This condensed string literal specifies only the actions and controls along with the qubit wires they are applied to. For example, a Toffoli gate in a 5-qubit circuit, with controls on the second and fourth wires and the action on the fifth wire, can be created with the call `cu_gate(sigma_x(1),` *"c2c4x5", 5*`)`. As implied by this example, n is the total number of qubits in the circuit that the gate is applied to. n must be a complex number with no imaginary component. n must also be within $10E-5$ of an integer value (e.g. 9.99999, 1.00001, and 3 are valid, but 4.5 is not), and values that are within this threshold are rounded to the nearest integer (e.g. 9.99999 is interpreted as 10). In addition, n must be at least 1 after rounding. More examples can be found in the examples/ directory and include hadder_bf1.qpro and rc_adder1.qpro, among others.
- `cutoff_val` If the cutoff value is set, any portion of all QuIDD element values that is less than the cutoff value will be rounded. For example, `cutoff_val =` $1e-15$ will cause all subsequently created QuIDD element values to be rounded at the 15th decimal place. By default, the cutoff value is not set and no rounding occurs. If the cutoff value is set by the user, it can be reset to the default (i.e. no rounding) by assigning 0 to `cutoff_val`.
- `dump_dot(`*"...", "...", a*`)` outputs the *dot* form of the graphical QuIDD representation of the matrix/vector a to a file specified by the second argument. The first argument is the name that will appear at the top of the QuIDD image. dot is a simple scripting language supported in the Graphviz package[1] Once the dot file is generated, dot can be run from the command line to produce a PostScript image of the QuIDD representation as such:
  `dot -Tps filename.dot -o filename.ps`
  dot can generate other graphical file formats as well. Consult Graphviz for

[1] Graphviz can be obtained at `http://www.graphviz.org/`.

more details. A simple example is contained in the examples/dot subdirectory.

- `echo("...")` prints the string literal passed as an argument to standard output. Putting one or more semicolons after **`echo`** does not suppress its output. **`echo`** has no return value, so it cannot be used in expressions.
- **`else`** is a program flow control construct that is part of an "if-elseif-else" control block sequence. Its meaning is the same as in just about any other language. Only one **`else`** may optionally appear in an "if-elseif-else" block, and it must appear only at the end of the block. If an **`else`** block is used, its body (a sequence of zero or more expressions and/or control blocks to be executed) must be terminated by an **`end`** even if the body is empty. The body following **`else`** is executed when the preceding **`if`** and **`elseif`** conditions evaluate to "false" (i.e. a complex numbered value of zero).
- **`elseif`** is a program flow control construct that is part of an "if-elseif-else" control block sequence. Its meaning is the same as in just about any other language. It contains a condition which is an expression enclosed in parentheses. Zero or more **`elseif`**'s may appear in an "if-elseif-else" block, but the first **`elseif`** must appear after an **`if`**, and the last **`elseif`** must appear before an optional **`else`**. If no **`else`** appears after an **`elseif`**, the body of the **`elseif`** (a sequence of zero or more expressions and/or control blocks to be executed) must be terminated by an **`end`** even if the body is empty. The condition determines whether or not the statements in the body are executed. The body of the **`elseif`** is executed when the following two conditions are met: 1.) the preceding **`if`** and **`elseif`** conditions evaluate to "false" (i.e. a complex numbered value of zero), and 2.) the **`elseif`** condition evaluates to "true" (i.e. any non-zero complex numbered value).
- **`end`** keyword that signifies the end of a program flow control construct. In other words, **`end`** should be used to denote the end of "if-elseif-else" and "while" blocks.
- **`exp`**(n) returns e^n. If n is a matrix, then it returns a matrix containing the element-wise computation of e^k where k is an element from n.
- **`expm`**(n) returns e^n, where n is a matrix. This is standard matrix exponentiation and is approximated by a finitely bounded Taylor series. In the current version of the QuIDDPro simulator, you may only apply **`expm`** to a matrix n whose dimensions do not exceed 8×8 for efficiency reasons. Future versions may support larger dimensional arguments, but it is unlikely that larger dimensional arguments will be needed for most quantum-mechanics applications. If n is a complex number, then it returns e^n.
- **`eye`**(n) returns an $n \times n$ identity matrix. If you only need an identity matrix whose dimensions are a power of 2 in size (e.g. for k-qubit identity gate matrices) then use $identity(k)$ instead (see below), which runs slightly faster. n must be a complex number with no imaginary component. n must also be within $10E-5$ of an integer value (e.g. 9.99999, 1.00001, and 3 are valid, but 4.5 is not), and values that are within this threshold are rounded to the nearest integer (e.g. 9.99999 is interpreted as 10). In addition, n

must be at least 1 after rounding. Always use `eye` or `identity` instead of defining identity matrices explicitly because they are asymptotically faster and use asymptotically less memory.

- `for` is a C-style looping construct. It involves a counter variable, a termination condition, and an incrementing expression, separated by *commas*.
- `fredkin()` returns a Fredkin gate matrix.
- `function` *var_name* = *func_name*($n_1, n_2, n_3, \ldots$) defines a function body. This definition should exist in a file by itself with a filename that matches *func_name* appended by the ".qpro" or ".qp" extensions. *var_name* is the name of the variable that contains the return value. n_1, n_2, and n_3 are function parameters that can be of any type. There is no constraint on the number of parameters. Also note that passing QuIDD arguments and QuIDD return values only requires $O(1)$ memory since only a single pointer to the head of a QuIDD needs to be passed. Arguments passed as parameters to functions are not modified by the function (i.e. pass-by-value is always used). Following the return value/function name line, the script code comprising the function body should appear. See Section A.2 for more details.
- `gen_amp_damp(`*d, p, n, a*`)` performs generalized amplitude dampening [61, p. 382]. Here a is a density matrix (it must be square and have dimensions that are a power of 2 in size) on which dampening is to be performed. a is not modified, but the result of dampening applied to a is returned. d is the dampening parameter and must be a complex number with no imaginary component. p is the probability parameter and must also be a complex number with no imaginary component. d and p must each be in the range $[0, 1]$. n is the qubit wire number that dampening is to be applied to. This wire number is only conceptual and can alternatively be thought of as the nth quantum state in the density matrix (see *cu_gate* for a more detailed description of wire numbers and steaneX.qpro and steaneZ.qpro under examples/nist/ for examples). n must be a complex number with no imaginary component. n must also be within $10E-5$ of an integer value (e.g. 9.99999, 1.00001, and 3 are valid, but 4.5 is not), and values that are within this threshold are rounded to the nearest integer (e.g. 9.99999 is interpreted as 10). In addition, n must be at least 1 after rounding.
- `hadamard(`n`)` or `H(`n`)` returns an n-qubit Hadamard gate matrix. n must be a complex number with no imaginary component. n must also be within $10E-5$ of an integer value (e.g. 9.99999, 1.00001, and 3 are valid, but 4.5 is not), and values that are within this threshold are rounded to the nearest integer (e.g. 9.99999 is interpreted as 10). In addition, n must be at least 1 after rounding. Always use this function instead of explicitly defining your own Hadamard matrix. This function is asymptotically faster and uses asymptotically less memory than defining the matrix explicitly.
- `i` is a variable that is preset to the value $0 + 1i$. It can be overwritten at runtime by the user.

- `identity(`n`)` returns an n-qubit identity gate matrix. n must be a complex number with no imaginary component. n must also be within $10E-5$ of an integer value (e.g. 9.99999, 1.00001, and 3 are valid, but 4.5 is not), and values that are within this threshold are rounded to the nearest integer (e.g. 9.99999 is interpreted as 10). In addition, n must be at least 1 after rounding. Always use this function instead of explicitly defining your own identity matrix. This function is asymptotically faster and uses asymptotically less memory than defining the matrix explicitly. Also see the *eye* function.
- `if` is a program flow control construct that is part of an "if-elseif-else" control block sequence. Its meaning is the same as in just about any other language. It contains a condition which is an expression enclosed in parentheses. An "if-elseif-else" block must be started by a single `if`, but "if-elseif-else" blocks can be nested within other "if-elseif-else" blocks (nesting with "while" blocks is also allowed). An `if` must be followed by a body of zero or more expressions and/or control blocks, and this body must be terminated by either an `elseif`, an `else`, or an `end`, even if the body is empty. The condition determines whether or not the statements in the body are executed. The body is executed once if the condition evaluates to "true" (i.e. any non-zero complex numbered value). Otherwise if the condition evaluates to "false" (i.e. a complex numbered value of zero), the body is not executed.
- `kron(`n, k`)` returns the tensor (Kronecker) product of the matrix expressions n and k. If n and k are complex numbers, then they are multiplied together.
- `lambda(`a, "...."`)` is an alternative name for the function `cu_gate`.
- `measure(`n, a`)` performs deterministic measurement on the nth qubit in the density matrix a. In other words, all off-diagonal correlations corresponding to the qubit being measured are zeroed out, and the resultant density matrix is returned (for probabilistic measurement of a qubit in a density matrix that returns a 1 or 0, see `pmeasure`). a must be square and have dimensions that are a power of 2 in size. a is not modified, but the result of measurement applied to a is returned. n is the qubit wire number that measurement is to be applied to. This wire number is only conceptual and can alternatively be thought of as the nth quantum state in the density matrix (see *cu_gate* for a more detailed description of wire numbers and steaneX.qpro and steaneZ.qpro under examples/nist/ for examples). n must be a complex number with no imaginary component. n must also be within $10E-5$ of an integer value (e.g. 9.99999, 1.00001, and 3 are valid, but 4.5 is not), and values that are within this threshold are rounded to the nearest integer (e.g. 9.99999 is interpreted as 10). In addition, n must be at least 1 after rounding.
- `measure_sv(`n, a`)` perform probabilistic measurement on qubit n. A state vector is returned which represents the state vector a as modified by the measurement result and its associated norm. If the measurement

result and the associated norm have already been computed with a previous call to `pmeasure_norm_sv`, then `measure_sv` can be called with the alternative syntax `measure_sv(`*n, a, res, norm*`)`. *res* and *norm* denote the precomputed measurement result and associated norm, respectively. Since a must be a state vector, one of the dimensions must be 1, and the other dimension must be a power of 2. a is not modified by this function. n must be a complex number with no imaginary component. n must also be within $10E-5$ of an integer value (e.g. 9.99999, 1.00001, and 3 are valid, but 4.5 is not), and values that are within this threshold are rounded to the nearest integer (e.g. 9.99999 is interpreted as 10). In addition, n must be at least 1 after rounding. *res* must have the value 0 or 1 to within the rounding threshold. *norm* should be a valid norm of a state vector.

- `measure_sv0(`*n, a*`)` returns the probability of measuring qubit n as a 0 in state vector a (for probabilistic measurement of a qubit in a state vector that returns a 1 or 0, see `pmeasure_sv`). Since a must be a state vector, one of the dimensions must be 1, and the other dimension must be a power of 2. a is not modified by this function. n must be a complex number with no imaginary component. n must also be within $10E-5$ of an integer value (e.g. 9.99999, 1.00001, and 3 are valid, but 4.5 is not), and values that are within this threshold are rounded to the nearest integer (e.g. 9.99999 is interpreted as 10). In addition, n must be at least 1 after rounding.
- `measure_sv1(`*n, a*`)` returns the probability of measuring qubit n as a 1 in state vector a (for probabilistic measurement of a qubit in a state vector that returns a 1 or 0, see `pmeasure_sv`). Since a must be a state vector, one of the dimensions must be 1, and the other dimension must be a power of 2. a is not modified by this function. n must be a complex number with no imaginary component. n must also be within $10E-5$ of an integer value (e.g. 9.99999, 1.00001, and 3 are valid, but 4.5 is not), and values that are within this threshold are rounded to the nearest integer (e.g. 9.99999 is interpreted as 10). In addition, n must be at least 1 after rounding.
- `norm(`*a*`)` returns the norm of a state vector or complex number a. Since a must be a state vector, one of the dimensions must be 1, and the other dimension must be a power of 2.
- `output_prec` denotes the output precision. When assigned a non-negative integer value, it specifies how many digits should be output to the screen. Any digits which exceed this number are rounded. For example, `output_prec = 3` will cause 1/3 to output 0.333 to the screen. Note that the internal precision of any numbers and variables are unaffected. `output_prec` only affects the screen output precision. By default, the variable `output_prec` is not set, but the output precision is initially 6. Assigning a negative value to `output_prec` restores the default output precision. However, assigning a matrix to `output_prec` leaves the precision unchanged from its previous value.

- `pi` is a variable that is preset to the value of π to a large number of decimal places. It can be overwritten at runtime by the user.
- `pmeasure(`n`,` a`)` performs probabilistic measurement on the nth qubit in the density matrix a. The result returned is a 1 or 0 (for deterministic measurement of a qubit in a density matrix, see **`measure`**). a must be square and have dimensions that are a power of 2 in size. a is not modified by this function. n must be a complex number with no imaginary component. n must also be within $10E-5$ of an integer value (e.g. 9.99999, 1.00001, and 3 are valid, but 4.5 is not), and values that are within this threshold are rounded to the nearest integer (e.g. 9.99999 is interpreted as 10). In addition, n must be at least 1 after rounding.
- `pmeasure_norm_sv(`n`,` a`)` performs probabilistic measurement on the nth qubit in the state vector a. A 1×2 vector is returned containing a 1 or 0 for the measurement result (the first element) and the norm associated with the measurement result (the second element). Since a must be a state vector, one of the dimensions must be 1, and the other dimension must be a power of 2. a is not modified by this function. n must be a complex number with no imaginary component. n must also be within $10E-5$ of an integer value (e.g. 9.99999, 1.00001, and 3 are valid, but 4.5 is not), and values that are within this threshold are rounded to the nearest integer (e.g. 9.99999 is interpreted as 10). In addition, n must be at least 1 after rounding.
- `pmeasure_sv(`n`,` a`)` performs probabilistic measurement on the nth qubit in the state vector a. The result returned is a 1 or 0 (for deterministic measurement of a qubit in a state vector see **`measure_sv0`** and **`measure_sv1`**). Since a must be a state vector, one of the dimensions must be 1, and the other dimension must be a power of 2. a is not modified by this function. n must be a complex number with no imaginary component. n must also be within $10E-5$ of an integer value (e.g. 9.99999, 1.00001, and 3 are valid, but 4.5 is not), and values that are within this threshold are rounded to the nearest integer (e.g. 9.99999 is interpreted as 10). In addition, n must be at least 1 after rounding.
- `proj0(`n`)` returns an n-qubit $|0\rangle$ projector gate matrix (i.e. $|0\ldots0\rangle\langle0\ldots0|$, for n 0's). n must be a complex number with no imaginary component. n must also be within $10E-5$ of an integer value (e.g. 9.99999, 1.00001, and 3 are valid, but 4.5 is not), and values that are within this threshold are rounded to the nearest integer (e.g. 9.99999 is interpreted as 10). In addition, n must be at least 1 after rounding. Always use this function instead of explicitly defining your own $|0\rangle$ projector matrix. This function is asymptotically faster and uses asymptotically less memory than defining the matrix explicitly.
- `proj1(`n`)` returns an n-qubit $|1\rangle$ projector gate matrix (i.e. $|1\ldots1\rangle\langle1\ldots1|$, for n 1's). n must be a complex number with no imaginary component. n must also be within $10E-5$ of an integer value (e.g. 9.99999, 1.00001, and 3 are valid, but 4.5 is not), and values that are within this threshold

are rounded to the nearest integer (e.g. 9.99999 is interpreted as 10). In addition, n must be at least 1 after rounding. Always use this function instead of explicitly defining your own $|1\rangle$ projector matrix. This function is asymptotically faster and uses asymptotically less memory than defining the matrix explicitly.

- `projplus(`n`)` returns an n-qubit $|+\rangle$ projector gate matrix (i.e. $|+\ldots+\rangle\langle+\ldots+|$, for n +'s). n must be a complex number with no imaginary component. n must also be within $10E-5$ of an integer value (e.g. 9.99999, 1.00001, and 3 are valid, but 4.5 is not), and values that are within this threshold are rounded to the nearest integer (e.g. 9.99999 is interpreted as 10). In addition, n must be at least 1 after rounding. Always use this function instead of explicitly defining your own $|+\rangle$ projector matrix. This function is asymptotically faster and uses asymptotically less memory than defining the matrix explicitly.
- `ptrace(`n, a`)` performs the partial trace over the nth qubit in the density matrix a. a must be square and have dimensions that are a power of 2 in size. a is not modified, but the result of the partial trace applied to a is returned. n is the qubit wire number that is traced over. This wire number is only conceptual and can alternatively be thought of as the nth quantum state in the density matrix (see *cu_gate* for a more detailed description of wire numbers and steaneX.qpro and steaneZ.qpro under examples/nist/ for examples). n must be a complex number with no imaginary component. n must also be within $10E-5$ of an integer value (e.g. 9.99999, 1.00001, and 3 are valid, but 4.5 is not), and values that are within this threshold are rounded to the nearest integer (e.g. 9.99999 is interpreted as 10). In addition, n must be at least 1 after rounding.
- `px(`p, n, a`)` applies a probabilistic Pauli X gate matrix to the nth qubit in the density matrix a. a must be square and have dimensions that are a power of 2 in size. a is not modified, but the result of dampening applied to a is returned. p is the probability parameter and must be a complex number with no imaginary component. p must be in the range $[0, 1]$. n is the qubit wire number that the probabilistic X gate matrix is to be applied to. This wire number is only conceptual and can alternatively be thought of as the nth quantum state in the density matrix (see *cu_gate* for a more detailed description of wire numbers and steaneX.qpro and steaneZ.qpro under examples/nist/ for examples). n must be a complex number with no imaginary component. n must also be within $10E-5$ of an integer value (e.g. 9.99999, 1.00001, and 3 are valid, but 4.5 is not), and values that are within this threshold are rounded to the nearest integer (e.g. 9.99999 is interpreted as 10). In addition, n must be at least 1 after rounding.
- `Px(`p, n, a`)` an alternative name for the function `px`.
- `py(`p, n, a`)` applies a probabilistic Pauli Y gate matrix to the nth qubit in the density matrix a. a must be square and have dimensions that are a power of 2 in size. a is not modified, but the result of dampening applied to a is returned. p is the probability parameter and must be a complex

number with no imaginary component. p must be in the range $[0, 1]$. n is the qubit wire number that the probabilistic Y gate matrix is to be applied to. This wire number is only conceptual and can alternatively be thought of as the nth quantum state in the density matrix (see *cu_gate* for a more detailed description of wire numbers and steaneX.qpro and steaneZ.qpro under examples/nist/ for examples). n must be a complex number with no imaginary component. n must also be within $10E-5$ of an integer value (e.g. 9.99999, 1.00001, and 3 are valid, but 4.5 is not), and values that are within this threshold are rounded to the nearest integer (e.g. 9.99999 is interpreted as 10). In addition, n must be at least 1 after rounding.

- `Py(`*p, n, a*`)` an alternative name for the function `py`.
- `pz(`*p, n, a*`)` applies a probabilistic Pauli Z gate matrix to the nth qubit in the density matrix a. a must be square and have dimensions that are a power of 2 in size. a is not modified, but the result of dampening applied to a is returned. p is the probability parameter and must be a complex number with no imaginary component. p must be in the range $[0, 1]$. n is the qubit wire number that the probabilistic Z gate matrix is to be applied to. This wire number is only conceptual and can alternatively be thought of as the nth quantum state in the density matrix (see *cu_gate* for a more detailed description of wire numbers and steaneX.qpro and steaneZ.qpro under examples/nist/ for examples). n must be a complex number with no imaginary component. n must also be within $10E-5$ of an integer value (e.g. 9.99999, 1.00001, and 3 are valid, but 4.5 is not), and values that are within this threshold are rounded to the nearest integer (e.g. 9.99999 is interpreted as 10). In addition, n must be at least 1 after rounding.
- `Pz(`*p, n, a*`)` an alternative name for the function `pz`.
- `qp_epsilon` This checks an internal cache when creating new QuIDD element values to see if those values have already been created. When checking the cache, QuIDDPro compares the new value to other values already in the cache up to epsilon. Specifically, a and b are considered equal if $abs(a-b) < epsilon*a$ and $abs(a-b) < epsilon*b$. Epsilon can be changed by assigning values to `qp_epsilon`. By default, the epsilon value is $1e-8$. Currently, the epsilon value is not always used when creating new QuIDD element values, but in future versions of QuIDDPro, the epsilon value will play a much greater role.
- `quidd_info(`*a*`)` prints information about an operator or state to standard output. This information includes the number of qubits represented (or acted upon), the dimensions of the explicit representation of the matrix, and the number of nodes in the QuIDD representation of the matrix. Note that the explicit matrix representation is not actually stored anywhere. a must be a valid operator, state vector, or density matrix.
- `r2` is a variable that is preset to the value of $\sqrt{2}$ to a large number of decimal places. It can be overwritten at runtime by the user.
- `r3` is a variable that is preset to the value of $\sqrt{3}$ to a large number of decimal places. It can be overwritten at runtime by the user.

- `rand(`n`)` returns a pseudo-random value between 0 and n. n can be any real value, including negative values.
- `round(`n`)` returns n with its real and imaginary parts rounded to the nearest integer. "Halfway" cases are rounded away from 0. Since there is no native integer type supported in QuIDDPro, round can be extremely helpful in ensuring that values which are supposed to be integer values are indeed integer values.
- `run "..."` executes all script code contained in the file specified by the argument. The run command may appear anywhere in a script except inside an explicit matrix. This command is ideal for declaring variables that may be re-used in multiple projects.
- `rx(`n, k`)` returns a k-qubit rotational Pauli X gate matrix given a real valued angle parameter n. n must be a complex number with no imaginary component. n must be in the range $[0, 1]$. k must be a complex number with no imaginary component. k must also be within $10E-5$ of an integer value (e.g. 9.99999, 1.00001, and 3 are valid, but 4.5 is not), and values that are within this threshold are rounded to the nearest integer (e.g. 9.99999 is interpreted as 10). In addition, k must be at least 1 after rounding.
- `Rx(`n, k`)` is an alternative name for the function `rx`.
- `ry(`n, k`)` returns a k-qubit rotational Pauli Y gate matrix given a real valued angle parameter n. n must be a complex number with no imaginary component. n must be in the range $[0, 1]$. k must be a complex number with no imaginary component. k must also be within $10E-5$ of an integer value (e.g. 9.99999, 1.00001, and 3 are valid, but 4.5 is not), and values that are within this threshold are rounded to the nearest integer (e.g. 9.99999 is interpreted as 10). In addition, k must be at least 1 after rounding.
- `Ry(`n, k`)` is an alternative name for the function `ry`.
- `rz(`n, k`)` returns a k-qubit rotational Pauli Z gate matrix given a real valued angle parameter n. n must be a complex number with no imaginary component. n must be in the range $[0, 1]$. k must be a complex number with no imaginary component. k must also be within $10E-5$ of an integer value (e.g. 9.99999, 1.00001, and 3 are valid, but 4.5 is not), and values that are within this threshold are rounded to the nearest integer (e.g. 9.99999 is interpreted as 10). In addition, k must be at least 1 after rounding.
- `Rz(`n, k`)` is an alternative name for the function `rz`.
- `sigma_x(`n`)` returns an n-qubit Pauli X gate matrix. n must be a complex number with no imaginary component. n must also be within $10E - 5$ of an integer value (e.g. 9.99999, 1.00001, and 3 are valid, but 4.5 is not), and values that are within this threshold are rounded to the nearest integer (e.g. 9.99999 is interpreted as 10). In addition, n must be at least 1 after rounding. Always use this function instead of explicitly defining your own X matrix. This function is asymptotically faster and uses asymptotically less memory than defining the matrix explicitly.
- `sigma_y(`n`)` returns an n-qubit Pauli Y gate matrix. n must be a complex number with no imaginary component. n must also be within $10E - 5$ of

an integer value (e.g. 9.99999, 1.00001, and 3 are valid, but 4.5 is not), and values that are within this threshold are rounded to the nearest integer (e.g. 9.99999 is interpreted as 10). In addition, n must be at least 1 after rounding. Always use this function instead of explicitly defining your own Y matrix. This function is asymptotically faster and uses asymptotically less memory than defining the matrix explicitly.

- `sigma_z(`n`)` returns an n-qubit Pauli Z gate matrix. n must be a complex number with no imaginary component. n must also be within $10E-5$ of an integer value (e.g. 9.99999, 1.00001, and 3 are valid, but 4.5 is not), and values that are within this threshold are rounded to the nearest integer (e.g. 9.99999 is interpreted as 10). In addition, n must be at least 1 after rounding. Always use this function instead of explicitly defining your own Z matrix. This function is asymptotically faster and uses asymptotically less memory than defining the matrix explicitly.
- `sin(`n`)` returns sine of the expression n passed as an argument. If n is a matrix, it returns a matrix containing the element-wise sine of n.
- `tan(`n`)` returns the tangent of the expression n passed as an argument. If n is a matrix, it returns a matrix containing the element-wise sine of n.
- `sqrt(`n`)` returns the square root of the expression n passed as an argument. If n is a matrix, it returns a matrix containing the element-wise square root of n.
- `swap(`n, k, a`)` returns the vector resulting from swapping qubits n and k in the state vector a. This function swaps qubits *much more quickly* than swapping using CNOT and Hadamard gates. Since a must be a state vector, one of the dimensions must be 1, and the other dimension must be a power of 2. a is not modified by this function. n and k must be complex numbers with no imaginary components. n and k must also be within $10E-5$ of an integer value (e.g. 9.99999, 1.00001, and 3 are valid, but 4.5 is not), and values that are within this threshold are rounded to the nearest integer (e.g. 9.99999 is interpreted as 10). n and k must also be at least 1 after rounding.
- `tic` starts a timer and also starts to record the peak memory usage from the point `tic` is called. `tic` has no return value, so it cannot be used in expressions. The timer only records time spent and memory used while running code. Thus, in the case of interactive mode, the timer will not be recording time spent nor memory used while at an idle prompt.
- `toc` stops a timer started by a previous `tic` or `toc` command. It outputs to standard output the time that has elapsed (i.e. time spent running code), the number of gates applied, the average runtime per gate, and memory that was used (peak memory) since the last `tic` or `toc` command. It also outputs the base memory which is the memory used in initializing the simulator and reading the input code. Base memory should be interpreted as a one-time initialization cost of the simulator and should not be considered when measuring performance. Operations that are recorded

as applied gates include matrix multiplication, `gen_amp_damp`, `measure`, `measure_sv`, `Px`, `Py`, and `Pz`.

- `toffoli("...")` returns a 3-qubit Toffoli gate matrix. This is a faster, specialized version of `cu_gate`. If a controlled gate matrix with different numbers of controls/targets and/or a different action (U operator) is desired, then use the more general `cu_gate` function. The string argument uses the same syntax as that of `cu_gate`. However, `toffoli` only accepts the strings *'ccx'*, *'cxc'*, and *'xcc'*, since these are the only valid Toffoli specifications. For example, `toffoli(`*'ccx'*`)` produces a Toffoli gate matrix with the controls on the "top" two wires and the action (X operator) on the "bottom" wire. For a discussion of how the concept of wires relates to creating controlled gate matrices, see `cu_gate`.
- `while` is a program flow control construct that allows multiple iterations of a body of code ("looping"). Its meaning is the same as in just about any other language. It contains a condition which is an expression enclosed in parentheses. A "while" block must be started by a single `while`, but "while" blocks can be nested within other "while" blocks (nesting with "if-elseif-else" blocks is also allowed). A `while` must be followed by a body of zero or more expressions and/or control blocks, and this body must be terminated by an `end`, even if the body is empty. The condition determines whether or not the statements in the body are executed. As long as the condition evaluates to "true" (i.e. any non-zero complex numbered value), the body is iteratively executed. The iterations stop when the condition becomes "false" (i.e. a complex numbered value of zero). The condition is checked once prior to executing each iteration of the body. `for` loops are also implemented with the counter variable, termination condition, and incrementing expression separated by commas.
- `zeros(`n, k`)` returns an $n \times k$ matrix of all 0's. x and y must be complex numbers with no imaginary components. n and k must be complex numbers with no imaginary component. n and k must also each be within $10E-5$ of an integer value (e.g. 9.99999, 1.00001, and 3 are valid, but 4.5 is not), and values that are within this threshold are rounded to the nearest integer (e.g. 9.99999 is interpreted as 10). In addition, n and k must each be at least 1 after rounding. Always use *zeros* instead of defining zero matrices explicitly because it is asymptotically faster and uses asymptotically less memory.

Appendix B

QuIDDPro Examples

This appendix provides sample QuIDDPro implementations of several well-known quantum circuits. These codes can be used with QuIDDPro software as explained in Section A.1. Section B.1 below contains small examples which create three well-known quantum states. Sections B.2 and B.3 offer larger examples which implement Grover's quantum search algorithm [38] and Shor's quantum integer factoring algorithm [82], respectively. A variety of other QuIDDPro scripts are supplied with the QuIDDPro distribution available at `http://vlsicad.eecs.umich.edu/Quantum/qp/`

B.1 Well-known Quantum States

This section contains QuIDDPro code which implements the cat (GHZ) state, the W state, and the equal superposition state. These examples illustrate how the language can be used to produce code that is as compact as the formal definition of such states.

Cat State

The cat state is an n-qubit generalization of the EPR pair and is defined as $|\psi_{\text{cat}}\rangle = (|00\ldots0\rangle + |11\ldots1\rangle)/\sqrt{2}$. A QuIDDPro function which creates this state given the number of qubits n is listed below.

```
function |cs:> = create_cat_state(n)
   |cs:> = (|0:>_n + |2^n - 1:>)/sqrt(2);
```

W State

The W state is an n-qubit state defined as $|\psi_{\text{W}}\rangle = (|10\ldots0\rangle + |01\ldots0\rangle + \ldots + |00\ldots1\rangle)/\sqrt{n}$. A QuIDDPro function which creates this state given the number of qubits n is given below.

```
function |ws:> = create_w_state(n)
   |ws:> = |1:>_n;
   for (j = 1, j < n, j++)
      |ws:> += |2^j:>;
   end
   |ws:> /= sqrt(n);
```

Equal Superposition State

The equal superposition state is an n-qubit state which represents all possible 2^n measurement outcomes with equal probability. It is defined as $\frac{1}{\sqrt{2^n}}\sum_{j=0}^{2^n-1}|j\rangle$ and can be created with Hadamard gates. A QuIDDPro function which creates this state given the number of qubits n is provided below.

```
function |sps:> = create_equal_superposition(n)
   |sps:> = H(n)*|0:>_n;
```

B.2 Grover's Search Algorithm

This section demonstrates how Grover's quantum search algorithm can be implemented in QuIDDPro. The first function provided below takes as arguments the number of qubits n, the oracle defining the search criteria, and an estimated number of items in the database which match the search criteria. This function returns the integer representation of the index measured of an item in the database (the left-most qubit in the state is most-significant). One ancillary qubit is used in conjunction with the oracle. As noted in the last appendix, assignment of QuIDDs only requires $O(1)$ time, which means that passing the oracle QuIDD to this function involves very little overhead. Another function is included later in this section which creates an oracle that searches for the last item in the database (the item with index $|11\ldots1\rangle$) which can be used in conjunction with the function implementing Grover's algorithm.

```
function index = grover_search(n, oracle, matches)
   |state:> = H(n)*|1:>_(n + 1);
   grover_op = H(n)*cps(n)*H(n)*oracle;
   # Compute the optimal number of Grover iterations.
   N = 2^n;
   x = sqrt(matches/N);
   theta = atan(x/sqrt(1 - x^2));
   num_iterations = pi/4/theta;
   # Perform the Grover iterations.
   for (g = 0, g < num_iterations, g++)
      |state:> = grover_op*|state:>;
```

```
   end
   # Measure an index.
   index = 0;
   for (q = 1, q <= n, q++)
      if (pmeasure_sv(q, |state:>))
         index += 2^(n - q);
      end
   end

function oracle = create_last_item_oracle(n)
   oracle_spec = ''x'';
   for (j = 0, j < n, j++)
      oracle_spec = ''c'' + oracle_spec;
   end
   oracle = cu_gate(sigma_x(1), oracle_spec);
```

B.3 Shor's Integer Factoring Algorithm

This section demonstrates a possible implementation of the main portion of Shor's algorithm. Given an integer N and its size in bits n, the following function uses quantum order-finding to find a non-trivial factor of N. Order-finding solves the problem of determining r such that $a^r \ mod\ N = 1$. For the purposes of factoring, a may be chosen randomly from the range $(1..N)$, and in the following function it is simply passed as an argument. Quantum modular exponentiation is used to compute all possible values for x and $a^x \ mod\ N$ simultaneously. Following this step, the inverse QFT is applied to increase the probability of measuring qubit values for which the state representation of $a^x \ mod\ N$ encodes the value 1 in binary. The value for x that is entangled with this part of the state is r. Classical post-processing is shown at the end, which makes use of the greatest common divisor algorithm. Not shown are functions implementing quantum modular exponentiation and the inverse QFT, each of which can be implemented in a variety of different ways [93, 92, 33].

```
function factor = shor_factor(N, a, n)
   if (N{1} == 0)
      factor = 2;
   else
      # Put the exponent state into an equal superposition.
      |x:> = H(n)*|0:>_n;
      |mod:> = |1:>_n;
      # Compute modular exponentiation and the inverse QFT.
      |res:> = mod_exp(|x:>, |mod:>, N, a, n);
      |res:> = inv_qft(|res:>, n);
      # Measure the exponent qubits.
```

```
    r = 0;
    for (q = 1, q <= n; q++)
       if (pmeasure_sv(q, |res:>))
          r += 2^(n - q);
       end
    end
    # Check if r can be used to calculate a factor.
    if ((r{1} == 0) && (rem(a^(r/2), N) != 1))
       cand_fac1 = gcd(a^(r/2) - 1, N);
       cand_fac2 = gcd(a^(r/2) + 1, N);
       if (rem(N, cand_fac1) == 0)
          factor = cand_fac1;
       elseif (rem(N, cand_fac2) == 0)
          factor = cand_fac2;
       else
          factor = -1;
       end
    else
       factor = -1;
    end
 end
```

References

1. Aaronson S, Gottesman D (2004) "Improved Simulation of Stabilizer Circuits," Phys Rev A 70:052328
2. Aaronson S (2003) "The CHP simulator," `http://www.scottaaronson.com/chp/`
3. Abdollahi A, Pedram M (2006) "Analysis and Synthesis of Quantum Circuits by Using Quantum Decision Diagrams," In Proc. of Design, Automation, and Test in Europe 317-322
4. Aharonov D, Landau Z, Makowsky J (2006) "The Quantum FFT can be Classically Simulated," quant-ph/0611156
5. Anders S, Briegel HJ (2006) "Fast Simulation of Stabilizer Circuits Using a Graph State Representation," Phys Rev A 73:022334
6. Bahar RI, et al. (1997) "Algebraic Decision Diagrams and their Applications," J of Formal Methods in System Design 10(2/3):171-206
7. Barenco A, et al. (1995) "Elementary Gates for Quantum Computation," Phys Rev A 52:3457-3467
8. Bennett CH (1992) "Quantum Cryptography Using Any Two Nonorthogonal States," Phys Rev Lett 68:3121-3124
9. Bennett CH, Brassard G (1984) "Quantum Cryptography: Public Key Distribution and Coin Tossing," In Proc. of IEEE Intl. Conf. on Computers, Systems, and Signal Processing 175-179
10. Bennett CH, Brassard G, Crépeau C, Jozsa R, Peres A, Wootters WK (1993) "Teleporting an Unknown Quantum State via Dual Classical and Einstein-Podolsky-Rosen Channels," Phys Rev Lett 70:1895
11. Berman GP, Doolen GD, López GV, Tsifrinovich VI (2000) "Simulations of Quantum-logic Operations in a Quantum Computer with a Large Number of Qubits," Phys Rev A 61:062305
12. Berman GP, et al. (2003) "Analytic Solutions for Quantum Logic Gates and Modeling Pulse Errors in a Quantum Computer with a Heisenberg Interaction," International J of Quantum Information 2(2):171-182
13. Berman GP, López GV, Tsifrinovich VI (2002) "Teleportation in a Nuclear Spin Quantum Computer," Phys Rev A 66:042312
14. Black PE, et al. (2003) "Quantum Compiling and Simulation," `http://hissa.nist.gov/~black/Quantum/`

15. Blondel VD, Jeandel E, Koiran P, Portier N (2005) "Decidable and Undecidable Problems about Quantum Automata," SIAM Journal on Computing 34:1464-1473
16. Boghosian BM, Taylor W (1998) "Simulating Quantum Mechanics on a Quantum Computer," Physica D 120:30-42
17. Boyer M, Brassard G, Hoeyer P, Tapp A (1998) "Tight Bounds on Quantum Searching," Fortsch Phys 46:493-506
18. Brennen GK (2002) "Distant Entanglement with Nearest Neighbor Interactions," quant-ph/0206199
19. Briegel HJ, Raussendorf R (2001) "Persistent Entanglement in Arrays of Interacting Particles," Phys Rev Lett 86:910-913
20. Bryant R (1986) "Graph-based Algorithms for Boolean Function Manipulation," IEEE Trans on Computers C35:677-691
21. Butscher B, Weimer H (2003) "Libquantum: the C Library for Quantum Computing," `http://www.enyo.de/libquantum/`
22. Celardo GL, Pineda C, Znidaric M (2003) "Stability of Quantum Fourier Transformation on Ising Quantum Computer," quant-ph/0310163
23. Chiaverini J, et al (2004) "Realization of Quantum Error Correction," Nature 432:602-605
24. Childs AM, Haselgrove HL, Nielsen MA (2003) "Lower Bounds on the Complexity of Simulating Quantum Gates," Phys Rev A 68:052311
25. Clarke E, et al. (1996) "Multi-terminal Binary Decision Diagrams and Hybrid Decision Diagrams," In: Sasao T, Fujita M (eds) Representations of discrete functions. Kluwer Academic Publishers, Norwell Dordrecht
26. Clarke E, Fujita M, McGeer PC, McMillan K, Yang J (1993) "Multi-Terminal Binary Decision Diagrams: An Efficient Data Structure for Matrix Representation," Intl Workshop on Logic and Synthesis 1-15
27. CNET News (2003) "Start-up Makes Quantum Leap in Cryptography,"
28. Cuccaro SA, Draper TG, Kutin SA, Moulton DP (2004) "A New Quantum Ripple-carry Addition Circuit," quant-ph/0410184
29. Draper TG, Kutin SA, Rains EM, Svore KM (2004) "A Logarithmic-depth Quantum Carry-lookahead Adder" quant-ph/0406142
30. Ekert A, Knight PL (1995) "Entangled Quantum Systems and the Schmidt Decomposition," Am J Phys 63(5):415-423
31. Ekert AK (1991) "Quantum Cryptography Based on Bell's Theorem," Phys Rev Lett 67:661-663
32. Fortnow L, Rogers J., "Complexity Limitations on Quantum Computation," Journal of Computer and System Sciences (1999), 59(2): 240-252
33. Fowler AG, Devitt SJ, Hollenberg LCL (2004) "Implementation of Shor's Algorithm on a Linear Nearest Neighbour Qubit Array," Quantum Information and Computation 4:237-251
34. Fowler AG, Hill CD, Hollenberg LCL (2004) "Quantum-error Correction on Linear-nearest-neighbor Qubit Arrays," Phys Rev A 69:042314
35. GNU MP (GMP): Arithmetic Without Limitations. http://www.swox.com/gmp/
36. Gottesman D (1998) "The Heisenberg Representation of Quantum Computers," Plenary speech at the 1998 Intl Conf on Group Theoretic Methods in Physics quant-ph/9807006
37. Greve D (1999) "QDD: a Quantum Computer Emulation Library," http://thegreves.com/david/QDD/qdd.html

38. Grover L (1997) “Quantum Mechanics Helps In Searching For A Needle In A Haystack,” Phys Rev Lett 79:325-328
39. Hachtel GD, Somenzi F (2006) “Logic Synthesis and Verification Algorithms,” Springer
40. Hayes JP (1993) “Introduction to Digital Logic Design,” Addison-Wesley
41. Hey AJG (ed) (2002) “Feynman and Computation: Exploring the Limits of Computers,” Westview Press, Boulder Cumnor Hill
42. Jaroszkiewicz G (2004) “Quantum Register Physics,” quant-ph/0409094
43. Jozsa R (2006) “On the Simulation of Quantum Circuits,” quant-ph/0603163
44. Jozsa R, Linden N (2002) “On the Role of Entanglement in Quantum Computational Speed-up,” quant-ph/0201143
45. Khodjasteh K, Lidar DA (2005) “Fault-tolerant Quantum Dynamical Decoupling,” Phys Rev Lett 95:180501
46. Kielpinski D, Meyer V, Rowe MA, Sackett CA, Itano WM, Monroe C, Wineland DJ (2001) “A Decoherence-free Quantum Memory Using Trapped Ions,” Science 291:1013-1015
47. Kielpinski D, Monroe C, Wineland DJ (2002) “Architecture for a Large-scale Ion-trap Quantum Computer,” Nature 417:709-711
48. Kitaev AY (1997) “Quantum Computations: Algorithms and Error Correction,” Russ Math Surv 52(6):1191-1249
49. Kitaev AY, Shen AH, Vyalyi MN (2002) “Classical and Quantum Computation,” American Mathematical Society, Graduate Studies in Mathematics, Providence
50. Ladd TD, Goldman JR, Yamaguchi F, Yamamoto Y (2002) “All-silicon Quantum Computer,” Phys Rev A 89:017901
51. Lee CY (1959) “Representation of Switching Circuits by Binary Decision Diagrams,” Bell System Tech J 38:985-999
52. Lidar DA, Wu LA (2003) “Quantum Computers and Decoherence: Exorcising the Demon from the Machine,” quant-ph/0302198
53. Lloyd S (1996) “Universal Quantum Simulators,” Science 273:1073-1078
54. Lu CY, Browne DE, Yang T, Pan JW (2007) “Demonstration of Shor’s Quantum Factoring Algorithm Using Photonic Qubits,” arXiv:0705.1684
55. Markov IL, Shi Y (2008) “Simulating Quantum Computation by Contracting Tensor Networks,” *SIAM Journal on Computing* 38(3): 963-981
56. Maslov D, Dueck G, Scott N “Reversible Logic Synthesis Benchmarks Page,” tt http://www.cs.uvic.ca/~dmaslov/
57. Metodiev T, Cross A, Thaker D, Brown K, Copsey D, Chong FT, Chuang IL (2004) “Preliminary Results on Simulating a Scalable Fault-tolerant Ion-trap System For Quantum Computation,” In Proc. of 3rd Workshop on Non-Silicon Computing
58. Miller DM, Thornton MA (2006) ”QMDD: A Decision Diagram Structure for Reversible and Quantum Circuits,” In Proc. of IEEE Intl. Symp. on Multiple-Valued Logic 30-35
59. Monroe C (2002) “Quantum Information Processing with Atoms and Photons,” Nature 416:238-246
60. Moore C, Crutchfield J (2000) ”Quantum Automata and Quantum Grammars,” Theoretical Computer Science 237:275-306
61. Nielsen MA, Chuang IL (2000) Quantum Computation and Quantum Information. Cambridge University Press, Cambridge

62. Obenland KM, Despain AM (1998) "A Parallel Quantum Computer Simulator," High Performance Computing
63. Ömer B QCL - a programming language for quantum computers http://tph.tuwien.ac.at/~oemer/qcl.html
64. Open Qubit Quantum Computing. lso arXiv:quant-ph/9804039v1. http://www.ennui.net/~quantum/
65. D. Perez-Garcia, F. Verstraete, J. I. Cirac, M. M. Wolf (2008) "PEPS as Unique Ground States of Local Hamiltonians," Quant. Inf. Comp. 8, 0650-0663
66. Petrosyan D, Kurizki G (2002) "Scalable Solid-state Quantum Processor Using Subradiant Two-atom States," Phys Rev Lett 89:207902
67. Prasad AK, Shende VV, Patel KN, Markov IL, Hayes JP (2006) "Data Structures and Algorithms for Simplifying Reversible Circuits," *ACM J. of Emerging Technologies in Computing* 2(4): 277-293
68. Protopopescu V, Perez R, D'Helon C, Schmulen J (2003) "Robust Control of Decoherence in Realistic One-qubit Quantum Gates," J Phys A: Math Gen 36:2175-2189
69. Stick D, Sterk JD, Monroe C (2007) "The Trap Technique," *IEEE Spectrum* 44(8): 36-43
70. QuIDDPro: high-performance quantum circuit simulation. `http://vlsicad.eecs.umich.edu/Quantum/qp/`
71. Ramalingam A, Nam GJ, Singh AK, Orshansky M, Nassif SR, Pan DZ (2006) "An Accurate Sparse Matrix Based Framework for Statistical Static Timing Analysis In Proc of *Intl Conf on Computer-Aided Design* 231-236
72. Sapatnekar S (2004) "Timing" 1st ed. Kluwer Academic Publishers, Norwell
73. Sasao T, Fujita M (eds) (1996) "Representations of discrete functions," Kluwer Academic Publishers, Norwell Dordrecht
74. Shankar R (1994) "Principles of Quantum Mechanics" 2nd ed. Springer Science+Business Media, New York
75. Shende VV, Bullock SS, Markov IL (2004) "Recognizing Small-circuit Structure in Two-qubit Operators," Phys Rev A 70:012310
76. Shende VV, Bullock SS, Markov IL (2006) "Synthesis of Quantum Logic Circuits," IEEE Trans on Computer-Aided Design 25:1000-1010
77. Shende VV, Markov IL (2005) "Quantum Circuits for Incompletely Specified Two-qubit Operators," Quantum Information and Computation 5(1):49-57
78. Shende VV, Markov IL (2009) "On the CNOT-cost of TOFFOLI gates," Quantum Information and Computation, 9(5-6):461-486
79. Shende VV, Prasad AK, Markov IL, Hayes JP (2003) "Synthesis of Reversible Logic Circuits," IEEE Trans on Computer-Aided Design 22(6):710-722
80. Shi Y (2003) "Both Toffoli and Controlled-NOT Need Little Help to do Universal Quantum Computation," Quantum Information and Computation, 3(1):84-92
81. Shi Y, Duan L, Vidal G (2006) "Classical Simulation of Quantum Many-body Systems with a Tree Tensor Network," Phys Rev A 74:022320
82. Shor PW (1997) "Polynomial-time Algorithms for Prime Factorization and Discrete Logarithms on a Quantum Computer," SIAM J of Computing 26:1484-1509
83. Somenzi F (1998) CUDD: CU decision diagram package. `http://vlsi.colorado.edu/~fabio/CUDD/cuddIntro.html`
84. Song G, Klappenecker A (2004) "Optimal Realizations of Simplified Toffoli Gates," Quantum Information and Computation 4:361-372

85. SPICE: Simulation Program with Integrated Circuit Emphasis. http://en.wikipedia.org/wiki/SPICE
86. Stanion RT, Bhattacharya D, Sechen C (1995) "An Efficient Method for Generating Exhaustive Test Sets," IEEE Trans on Computer-Aided Design 14:1516-1525
87. Steane AM (1996) "Error-correcting Codes in Quantum Theory," Phys Rev Lett 77:793
88. Steane AM (2001) "Quantum Computing and Error Correction," In: Gonis T, Turchi PEA (eds) Decoherence and its implications in quantum computation and information transfer. IOS Press, Amsterdam
89. Strang G (2005) Linear Algebra and Its Applications 4th ed. Brooks Cole
90. Tian L, Zoller P (2003) "Quantum Computing with Atomic Josephson Junction Arrays," quant-ph/0306085
91. Valiant LG (2001) "Quantum Computers that Can be Simulated Classically in Polynomial Time," In Proc of ACM Symp on Theory of Computing 114-123
92. Van Meter R, Itoh KM (2005) "Fast Quantum Modular Exponentiation," Phys Rev A 71:052320
93. Vedral V, Barenco A, Ekert A (1996) "Quantum Networks for Elementary Arithmetic Operations," Phys Rev A 54:147-153
94. Veldhuizen T (1998) "Blitz," In Proc 2nd Intl Symp on Computing in OO Parallel Environments `http://www.oonumerics.org/blitz/`
95. Viamontes GF, Markov IL, Hayes JP (2005) "Graph-based Simulation of Quantum Computation in the Density-matrix Representation," Quantum Information and Computation 5(2):113-130
96. Viamontes GF, Markov IL, Hayes JP (2005) "Is Quantum Search Practical?" Computing in Science and Engineering 7(4):22-30
97. Viamontes GF, Markov IL, Hayes JP (2004) "Graph-based Simulation of Quantum Computation in the Density Matrix Representation," In Proc of SPIE 5436:285-296
98. Viamontes GF, Markov IL, Hayes JP (2004) "High-performance QuIDD-based Simulation of Quantum Circuits," In Proc of the Design, Automation and Test in Europe Conference 2:1354-1355
99. Viamontes GF, Markov IL, Hayes JP (2003) "Improving Gate-level Simulation of Quantum Circuits," Quantum Information Processing 2(5):347-380
100. Viamontes GF, Markov IL, Hayes JP (2007) "Checking Equivalence of Quantum Circuits and States," In Proc of Intl Conf on Computer-Aided Design 69-74
101. Viamontes GF, Rajagopolan M, Markov IL, Hayes JP (2003) "Gate-level Simulation of Quantum Circuits," In Proc of the Asia South Pacific Design Automation Conf 295-301
102. Viamontes GF, Rajagopalan M, Markov IL, Hayes JP (2002) "Gate-Level Simulation of Quantum Circuits," In Proc of the 6th Intl Conference on Quantum Communication, Measurement, and Computing 311-314
103. Vandersypen LMK, Steffen M, Breyta G, Yannoni CS, Sherwood MH, Chuang IL (2001) "Experimental Realization of Shor's Quantum Factoring Algorithm Using Nuclear Magnetic Resonance," Nature 414:883-887
104. Vandersypen L (2007) "Dot-to-Dot Design," *IEEE Spectrum* 44(9): 42-47
105. Verstraete F, Garcia-Ripoll JJ, Cirac JI (2004) "Matrix Product Density Operators: Simulation of finite-T and dissipative systems," Phys Rev Lett, 93:207204

106. Vidal G (2003) "Efficient Classical Simulation of Slightly Entangled Quantum Computations," Phys Rev Lett 91:147902
107. Vidal G (2004) "Efficient Simulation of One-dimensional Quantum Many-Body Systems," Phys Rev Lett 93:040502
108. Viola L, Knill E, Lloyd S (1999) "Dynamical Decoupling of Open Quantum Systems," Phys Rev Lett 82:2417-2421
109. Viola L, Lloyd S (1998) "Dynamical Suppression of Decoherence in Two-state Quantum Systems," Phys Rev A 58:2733-2744
110. Viola L, Lloyd S, Knill E (1999) "Universal Control of Decoupled Quantum Systems," Phys Rev Lett 83:4888
111. Vrijen R, et. al (2000) "Electron-spin-resonance Transistors for Quantum Computing in Silicon-germanium Heterostructures," Phys Rev A 62:012306
112. Yoran N, Short AJ (2007) "Efficient Classical Simulation of the Approximate Quantum Fourier Transform," Phys Rev A 76:043221
113. Zwolak M, Vidal G (2004) "Mixed-state Dynamics in One-dimensional Quantum Lattice Systems: a Time-dependent Super-operator Renormalization algorithm," Phys Rev Lett 93:207205

Index

ADD, *see* algebraic decision diagram
algebraic decision diagram, 15, 71
ancillary qubit, 32, 40, 44
Apply, 13
asynchronous circuit, 17

bang-bang error correction, 140
BDD, 11, *see* binary decision diagram
Bell state, 30, 41
binary decision diagram, 11, 18
Bloch sphere, 35
Boolean predicate, 44
bra-ket, 24

CAD, *see* computer-aided design
cat state, 62, 177
CHP simulator, 54
circuit simulation, 7
classical
  circuits, 7
  communication, 2
  computing, 1
  control, 43
  equivalence checking, 10
  simulation, 8
  verification, 10
Clifford group, 50
clock cycle, 17
clock period, 17
CNOT gate, 34, 35, 39–41, 44
CNOT operator, 30, 50
collective dephasing, 144
commutator, 49
complex conjugate, 20, 21
computational basis, 24, 26, 27, 29
computer-aided design, 3, 153
conditional phase-shift, 44
control qubit, 30, 39, 44
control qubit, 30, 39, 44
controlled gate, 43
corrective pulse, 140
CUDD, 73

D flip-flop, 17
decision diagram, 18
decoherence, 27
decoherence error, 144
dense tensor decomposition, 66
density matrix, 31, 32, 37
digital circuit, 3, 7, 17, 32, 33, 40, 42, 153
Dirac model, 24
Dirac notation, 24, 28
DTED, *see* dense tensor decomposition

eigenvalue, 23
eigenvector, 23
electronic spin, 33
element-wise division, 123
element-wise multiplication, 77
environment qubit, 144
environmental noise, 37
EPR pair, 30, 31, 41, 53
equal superposition, 25, 26, 40, 44, 178
equivalence
  checking, classical, 10
  checking, quantum, 115
  global-phase, 116, 117
  relative-phase, 116, 122

error-correcting code, 140
exponential runtime, 3, 45

factoring, 1, 43, 45, 48
fanout, 40
finite-state automaton, 18
finite-state machine, 18, 42
FSA, *see* finite-state automaton
FSM, *see* finite-state machine
fundamental postulates, 24, 33, 37, 38

gallium arsenide, 2
gate
  $\pi/8$, 52
  Clifford, 50
  CNOT, *see* CNOT gate
  controlled, 37
  error model, 142
  Hadamard, *see* Hadamard gate
  library, 41
  modeling, 7
  NAND, *see* NAND gate
  NOT, 7, 34
  Pauli, 34
  phase, *see* phase gate
  rotation, *see* rotation operator
  SWAP, *see* SWAP gate
  Toffoli, *see* Toffoli gate
  XOR, *see* XOR gate
global phase, 35, 51
greatest common divisor, 45, 46
Grover's algorithm, 4, 43, 44, 85, 113, 119, 178

Hadamard gate, 36, 40, 44, 50
Hadamard operator, 26, 28, 30, 50
Hamiltonian circuit, 127
Heisenberg representation, 49
Hilbert space, 21, 23, 24, 29, 32

inner product, 21, 22, 24, 29
ion trap, 2, 42, 144

ket, *see* bra-ket

linear algebra, 19, 24, 32, 33
look-up table, 7, 8

MATLAB, 155
matrix
  adjoint, 23, 26, 37
  complex conjugate, 23
  Hermitian, 23
  identity, 22, 24, 34, 40
  inverse, 22, 23, 26
  Kronecker product, *see* tensor product
  multiplication, 21, 39
  operations, 21
  Pauli, 23, 50, 52
  projection, 22, 26
  square, 21–23
  tensor product, 22, 27, 28, 32
  trace, 21, 31
  transpose, 21, 23
  unitary, 23, 25, 26, 34, 35, 40, 142
matrix-vector product, 24–26
modular exponentiation, 46, 64
MTBDD, *see* multi-terminal binary decision diagram
multi-terminal binary decision diagram, 15, 71

NAND gate, 30
National Institute of Standards and Technology, 2
nearest-neighbor, 46
nearest-neighbor interaction, 140
NIST, *see* National Institute of Standards and Technology
no-cloning theorem, 25, 29, 32, 40
node-count equivalence check, 118
non-0 terminal merge, 125
non-resonant effects, 143
NP, 13, 15, 115
nuclear magnetic resonance, 42
nuclear spin, 33

operator, 25, 26, 28
oracle, 3, 44
orthogonal, 22, 29
outer product, 22

partial trace, 31, 32, 37, 38
particle qubit, 42
Pauli matrices, 23, 50, 52
phase
  dampening, 144
  estimation, 45

gate, 36, 41
operator, 50
photon polarization, 33
probability amplitude, 25, 26, 40, 44, 46
projective measurement, 26

QDD, *see* quantum decision diagram
QFA, *see* quantum finite automaton
QFT, *see* Quantum Fourier Transform
QMA-complete, 115
QMDD, *see* quantum multiple-valued decision diagram
quadratic runtime improvement, 43
quantum
algorithm, 1, 43
bit, *see* qubit
circuit, 2, 33, 36, 38, 40, 42–45, 47, 48, 153
circuit classes, 48, 49, 154
circuit simulation, 18, 31, 33, 47
communication, 2, 42
computation, 29, 39
computing, 1, 24–26, 32, 41, 42
decision diagram, 98
entanglement, 30, 32, 37, 38, 40, 45
equivalence, *see* equivalence
equivalence checking, 115
error correction, 36, 42, 51
factoring, 45
finite automaton, 18, 42, 43
flip-flop, 42
Fourier Transform, 4, 46, 64, 101
gate, 33, 38, 39, 42, 48
half-adder, 38, 39
measurement, 26, 31, 37, 40, 43, 46
mechanics, 1, 19, 21, 23, 24, 32
multiple-valued decision diagram, 96
noise, 27, 32
search, 5, 43, 44
simulation, 3, 47, 103
simulator, *see* simulator
states, 24, 26, 27, 29
superposition, 11, 24–26, 43, 45
teleportation, 31
quantum information decision diagram, 4, 5
complexity of operations, 80
density matrix benchmarks, 109
density matrix example, 104
density-matrix simulation, 103
dynamic tensor product, 139
global-phase equivalence check, 118
implementation insights, 133
inverse QFT representation, 100
linear-algebraic operations, 75
matrix multiplication, 76
measurement, 77
multiplication example, 72
multiplicative edge values, 92
numerical aspects, 78
outer product, 105
partial trace, 106
representation, 72
scalability, 80
Shor state representation, 120
simulating Grover's algorithm, 88, 113
special gate algorithms, 133
state-vector simulation, 71
tensor product, 75
variable ordering, 73
quantum-optical computation, 42
qubit, 1, 24, 25, 32, 33, 38
superposition, 26, 178
qubit-wise multiplication, 59
query complexity, 43
QuIDD, *see* quantum information decision diagram, 18
QuIDDPro, 4, 44, 71, 73, 109, 154, 155
examples, 177
language reference, 160

reduced density matrix, 31
reduced ordered binary decision diagram, 3, 12
remote
entanglement, 141
EPR pair, 126, 153
EPR-pair circuit, 141, 143
reversible circuit, 25, 38, 40, 41
RF pulse, 42
ROBDD, *see* reduced ordered binary decision diagram, 12–15
rotation operators, 35, 41

Schrödinger equation, 11
search, 1, 43
secure public key exchange, 31

sequential circuit, 17, 42
Shor's algorithm, 1, 43, 45, 46, 120, 179
simulation
  1-qubit gate algorithm, 134
  electrical, 7
  functional, 8
  Grover's algorithm, 85
  hybrid techniques, 153
  modular exponentiation, 64
  P-blocked, 61
  Pfaffian, 70
  quantum, 3, 4, 32, 47, 103
  quantum circuit, 33
  QFT, 64, 100
  qubit-wise, 59
  Shor's algorithm, 64, 98, 100
  slightly-entangled, 4, 66, 153
  stabilizer circuits, *see* stabilizer formalism, 54
  symbolic, 9, 10, 71, 103
  tensor networks, 64
  timing, 8, 9
  using look-up tables, 7
  Vidal's technique, 66
simulator, 4, 47, 54, 155, 177
Solovay-Kitaev theorem, 41
spectral representation, 24
stabilizer
  circuit, 4, 47, 49
  formalism, 5, 47
  gates, 36, 41
state fidelity, 148
stationary qubit, 42
stationary state, 42
surd, 79
SWAP gate, 34
symbolic arithmetic, 79
synchronization, 17, 42
systematic error, 143

target qubit, 30, 39
tensor networks, 4, 63, 153
time-dependent evolution, 34
timing analysis, 7–9
Toffoli gate, 30, 36, 39, 41, 116
Toffoli operator, 30
truth table, 8, 9

universal gate library, 30, 36, 41

very large-scale integration, 2
VLSI, *see* very large-scale integration

XOR gate, 30

Kolibri 360 GmbH
Ferdinand-Jühlke-Straße 7
99095 Erfurt, Deutschland
produktsicherheit@kolibri360.de